# Living Shorelines for Florida

# Living Shorelines for Florida

## A Practical Guide for Building Coastal Resilience

SAVANNA BARRY, VINCENT ENCOMIO,
MARJORIE SHROPSHIRE,
AND GINNY STIBOLT

University of Florida Press
GAINESVILLE

*Cover*: A living shoreline in Cedar Key, Florida. Photo by Mark Clark, UF/IFAS.

This publication was supported by the National Sea Grant College Program of the United States Department of Commerce, National Oceanic and Atmospheric Administration, Grant NA240ARX417C0144. The views expressed are those of the authors and do not necessarily reflect the views of these organizations. Additional information is available by contacting Florida Sea Grant, University of Florida, P.O. Box 110409, Gainesville, FL, 32611-0409, tel. (352) 392-2801, www.flseagrant.org.


Published in the United States of America
30 29 28 27 26 25    6 5 4 3 2 1
ISBN: 978-1-68340-550-4
DOI: https://doi.org/10.5744/9781683405504

Library of Congress Cataloging-in-Publication Data
Names: Barry, Savanna author | Encomio, Vincent author | Shropshire, Marjorie author | Stibolt, Ginny author
Title: Living shorelines for Florida : a practical guide for building coastal resilience / Savanna Barry, Vincent Encomio, Marjorie Shropshire, Ginny Stibolt.
Description: Gainesville : University of Florida Press, 2025. | Includes index. | Summary: "An important resource for anyone involved in managing waterfront property in Florida, this book explains the concept of living shorelines-nature-based coastal infrastructure and landscaping-and how to implement ecologically informed shoreline protection in the state"— Provided by publisher.
Identifiers: LCCN 2025019196 (print) | LCCN 2025019197 (ebook) | ISBN 9781683405504 hardback | ISBN 9781683405658 ebook | ISBN 9781683405580 pdf
Subjects: LCSH: Shore protection—Florida | Coastal zone management—Florida | Shorelines—Monitoring—Florida | Coast changes—Florida | Beach erosion—Florida | BISAC: NATURE / Ecosystems & Habitats / Coastal Regions & Shorelines | NATURE / Environmental Conservation & Protection
Classification: LCC TC224.F6 B37 2025 (print) | LCC TC224.F6 (ebook) | DDC 627/.5809759—dc23/eng/20250821
LC record available at https://lccn.loc.gov/2025019196
LC ebook record available at https://lccn.loc.gov/2025019197

UF PRESS
University of Florida Press
2046 NE Waldo Road
Suite 2100
Gainesville, FL 32609
http://upress.ufl.edu

GPSR EU Authorized Representative: Mare Nostrum Group B.V., Mauritskade 21D, 1091 GC Amsterdam, The Netherlands, gpsr@mare-nostrum.co.uk

# Contents

# Preface

**THE ORIGINS OF THIS** book trace back to discussions held years ago between Marjorie and Vincent on increasing interest in living shorelines among homeowners through art and design. Eventually, Marjorie pursued her MFA in design and visual communications at the University of Florida, and her thesis tackled the social barriers to living shoreline adoption through intentional co-design. We thank Marjorie's committee members in the Master of Fine Arts in Design and Visual Communications program—Professors Maria Rogal (committee chair), Jarred Elrod, Vincent Encomio, and Katerie Gladdys—for helping distill her concepts into the project that inspired this book. Several of the diagrams in this book are directly from Marjorie's thesis. Another source of inspiration came when Marjorie took Savanna's University of Florida Institute of Food and Agricultural Sciences (IFAS) Florida Master Naturalist Program (FMNP) course on Coastal Shoreline Restoration, which enhanced her knowledge of the estuarine habitats that living shorelines aim to restore. Marjorie's past work with Ginny coauthoring or illustrating books on native plants and sustainable gardening also helped in the conception of a book to educate a broader audience about living shorelines. And finally, the deep knowledge, professional expertise and on-the-ground experience Savanna and Vincent brought to this project turned the concept into reality and led to the book you now hold in your hands.

The information herein is the culmination of discussions and practice with many partners. We thank our Florida Sea Grant colleagues from the Living Shorelines Training for Marine Contractors instructional team: Armando Ubeda, Mandy Baily, Rick O'Connor, Carolyn Kovacs, and Mike Sipos. Other partnering agencies and institutions in this course are dedicated staff from the Florida Department of Environmental Protection, the US Army Corps of Engineers, the Florida Fish and Wildlife Conservation Commission, several local governments in Florida that assisted with instruction, and Tom Ries from the Ecosphere Restoration Institute. The teaching contributions of these organizations and engagement with the many past course participants helped refine how we address the different audiences that will use this book. We also thank Florida Sea Grant Director and Associate Extension Director, Drs. Sherry Larkin and Maia Patterson McGuire for their support of the book and its publication. We thank Dr. Mark Clark for his unmatched dedication to pioneering new living shoreline techniques, his generous mentorship of others in living shorelines practice, and his bottomless efforts to improve Florida's environment.

We also thank the many instructors of the Florida Master Naturalist Program course in Coastal Shoreline Restoration and Dr. Martin Main, the FMNP program leader, for teaching this valuable course, which oftentimes is the first introduction to estuarine habitats and living shorelines to new Florida residents. The FMNP is highly impactful in educating the public on the importance of natural Florida, and we highly recommend its courses to all Floridians. Cammie Donaldson from the Florida Association of Native Nurseries provided insightful comments on the crucial role the landscape industry can play in living shoreline practice. We've all talked to countless additional people to work through various issues on this complex topic, and we appreciate their help.

We also thank Restore America's Estuaries for hosting the biennial Living Shorelines Technology Transfer Conferences. These conferences have served as a crucial platform for the exchange of techniques, lessons learned, and best practices within the national living shorelines community. Much of the information presented at these conferences over the years has benefited Florida living shorelines practitioners and, in turn, has found its way into this book.

We are pleased that the University of Florida Press agreed to pursue this book with us and hired anonymous peer reviewers. We deeply appreciate their thoughtful and thorough comments and suggestions that have significantly improved the manuscript and our focus.

Special thanks to Janie Chan, our editor at the Press, for guiding us through the process. Janie was generous with her time and resources, and her ideas for coordinating the writing and editing were invaluable. And, of course, thanks to all the other people at the Press who have worked through the various stages in creating and marketing this book.

We thank all who gave us permission to use their images in this book. They represent your partnerships and impactful contributions to moving the needle forward on living shorelines.

And thank you, the reader, for finding this book and starting this journey towards a more resilient Florida.

Finally, we thank our families for their support and love.

# Introduction

**THE TERM "LIVING SHORELINES"** refers to a multifaceted approach to coastal management and shoreline stabilization that prioritizes natural coastal processes manifested through native coastal vegetation, oyster reefs, and other nature-based approaches over traditional coastal armoring structures. As human populations continue to grow and as coastal areas become increasingly vulnerable to the impacts of climate change, sea level rise, and storms, the need for effective shoreline management strategies has never been more pressing. Living shorelines allow us to manage our coastal environment in a cost-effective and environmentally sound manner while avoiding the harmful effects and high costs associated with traditional armoring.

## Living Shorelines—Why Do We Need More of Them?

In the United States, shoreline hardening is a widespread practice that dates back to the early twentieth century. Florida, with its extensive coastline and substantial property values, has been particularly affected by this trend. Florida's shorelines are increasingly experiencing accelerated erosion, primarily due to the loss of natural wetlands and the use of hardening structures such as seawalls and jetties that disrupt sediment transport and intensify erosion on adjacent shorelines. These developments have led to increased costs for property owners, reduced property values, and decreased ecological well-being.

Despite the growing acceptance of living shorelines as a preferred approach to coastal management, human activities have led to the widespread degradation of coastal ecosystems. In turn, this has led to the loss of the valuable functions that coastal ecosystems provide, such as buffering of erosional forces, fisheries production, and water quality improvement. In the United States, installation of armoring structures—such as seawalls, jetties, and revetments—is widespread. Shorelines are heavily degraded in coastal communities, particularly in fast-growing, highly urbanized states such as Florida. Many people are seeking more sustainable and resilient solutions to address the challenges posed by a changing climate. Living shorelines, as you will discover in this book, can be a big part of resilient solutions for the future.

As climate change becomes ever more evident, it has become paramount to pioneer innovative approaches to shoreline management that protect both human and environmental wellbeing, as the two are inextricably linked. Coastal ecosystems—such as mangrove forests, salt marshes, oyster reefs, and sand dunes—provide critical habitat for a wide range of plant and animal species while also serving as highly effective natural buffers against storms and sea level rise. Living shorelines, which prioritize ecological integrity and native species, hold promise as a key strategy for building resilience and promoting long-term sustainability in coastal areas.

**FIGURE 00.01** Saltmarsh grasses and oyster reefs are important components of many living shorelines. *UF/IFAS file photo by Tyler Jones.*

## The Purpose of This Book

This book provides a comprehensive exploration of the concept of living shorelines and their application in Florida, focusing on the key principles, best practices, and practical tips related to this multifaceted approach to coastal management. Through real-world examples, step-by-step guides, and insider insights from experts in the field, this book will equip readers with the knowledge and resources needed to design, implement, and maintain successful living shorelines projects.

## What's in This Book

This book is organized around the following key topics:

Chapters 1 and 2 cover background information on Florida's key coastal habitats and how human activities impact them—both negatively and positively.
Chapter 3 introduces living shorelines, their history, and effectiveness in comparison to shoreline armoring.
Chapters 4 and 5 describe aspects to consider before getting started, the types of living shoreline designs, and the building blocks of living shorelines.
Chapters 6 and 7 offer step-by-step guidance for planning, implementing, and maintaining your shoreline, including advice on selecting materials, plants, and construction methods for living shoreline projects.
Chapters 8 and 9 cover shoreline-friendly approaches ranging from building buffers to managing upland landscapes.

Throughout the book, case studies and practical insights from real-world living shorelines in Florida are included. There are also several helpful appendices and a glossary section for reference.

## Audiences for This Book

This book has been designed with a wide range of audiences in mind, including property owners, contractors, local governments, and other stakeholders, while recognizing that each of these groups has unique needs and challenges.

**Property Owners**: Property owners and homeowners looking for ways to protect their investments may be interested in exploring their options for shoreline management. Whether property owners are looking for guidance for a do-it-yourself solution or seeking insights into the process to better communicate with hired professionals about their goals, this book can help. The chapters provide detailed background information as well as step-by-step guidance on assessing site suitability, selecting materials, choosing native plant species, and navigating the regulatory process to build robust living shorelines.

**Contractors and Consultants**: A variety of professionals such as landscape architects, engineers, or project managers may be tasked with implementing living shorelines projects in coastal communities. This book can serve as a valuable tool for communicating with clients about living shorelines or as a training resource for junior colleagues who may be new to the process. With practical information on designing and building successful shorelines, accompanied by a variety of diagrams and case studies to serve as reference material, this book will serve as a handy resource for practitioners.

**Government and Regulatory Agencies**: Representatives of local government departments such as parks and recreation, public works, and environmental protection, as well as state or federal land management agencies, are increasingly looking to develop sustainable coastal protection strategies for their communities. This book can help public sector staff design and develop living shorelines and communicate about them with their constituents. Furthermore, permitting officials from local, state, and federal offices must evaluate a variety of coastal projects, including living shorelines, to determine if construction is allowable. This book explores the policy and regulatory frameworks necessary for implementing living shorelines projects from the perspective of a permit applicant, giving insights to regulators about how practitioners formulate their project plans. Moreover, this book presents best practices for living shoreline design, and it covers a wide array of plants, reef materials, and bank stabilization techniques that are important to understand from a regulatory perspective. Thus, this book can be used to help train new regulatory staff about living shorelines, and it can serve as a reference for experienced regulatory staff when evaluating permit applications for living shorelines.

**Nongovernmental Organizations (NGOs):** Environmental and conservation-oriented NGOs or nonprofits are likely the most interested in achieving the ecological benefits of living shorelines. NGOs that are new to, or even experienced in, habitat restoration and nature-based infrastructure will find this book useful as a practitioner's guide. Like the aforementioned audiences, NGOs will benefit, as practitioners, from the details in the book on planning, design, permitting, and installation of living shorelines. The variety of living shoreline designs and materials described in this book can serve as a baseline and, hopefully, as inspiration to develop newer designs, approaches, and sustainable materials that align with an NGO's mission of conservation and sustainability. This book will also serve as an effective outreach tool for the public, particularly for homeowners, in an NGO's mission to promote environmentally friendly practices. It can also serve as a useful reference for NGO staffers developing their own outreach materials related to living shorelines. Using the knowledge gained from this book, an NGO may act as a facilitator among clients, contractors, and government agencies to educate and promote living shorelines and other nature-based solutions.

## How to Use This Book

Within these four audiences, individuals may have varying degrees of knowledge, education, and experience with living shorelines. Feel free to focus on particular sections or to skip ahead to the chapters that are most relevant to the projects at hand or to the current stage of a living-shoreline practice. For people who are new to this subject or who are just learning about Florida's ecosystems and living shorelines, it will be helpful to read this book from beginning to end. This book has been designed to serve as a reference for additional living shoreline projects going forward. Please share what you learn and your experiences with others, because together we can help make Florida's coastal regions more resilient in a changing world.

# 1

# Introduction to Florida's Coastal Zones

**THERE ARE 1,350 MILES** (2,172.6 km) of coastline in Florida, and much of the state's distinctive character lies in the beauty of its coastline. So it is not surprising that more than 75 percent of Florida's population lives in or near these coastal zones. (See chapter 2 for a summary of how people have both positively and negatively affected Florida's coastal ecosystems over time.) This chapter covers the most important ecosystems in Florida's coastal zones and how they function.

## Ecosystems and Habitat

The Convention on Biological Diversity (1992) defines an ecosystem as "a complex of living organisms and the abiotic environment with which they interact in a specified location."[1] In other words, an ecosystem is a local network of interacting biotic (living organisms: plants, animals, fungi, bacteria, and others) and abiotic conditions (temperature, water, acidity, and minerals in the substrate) in an environment.

Ecosystems can be large or small, and biotic and abiotic factors in these ecosystems all work together, either directly or indirectly. A change in one factor, such as the acidity or salinity of water in an ecosystem, may change the types of plants that can grow in that ecosystem. When changes in the ecosystem occur, the organisms that depend on plants for food and shelter will have to adapt to those changes, relocate to another ecosystem, or die.

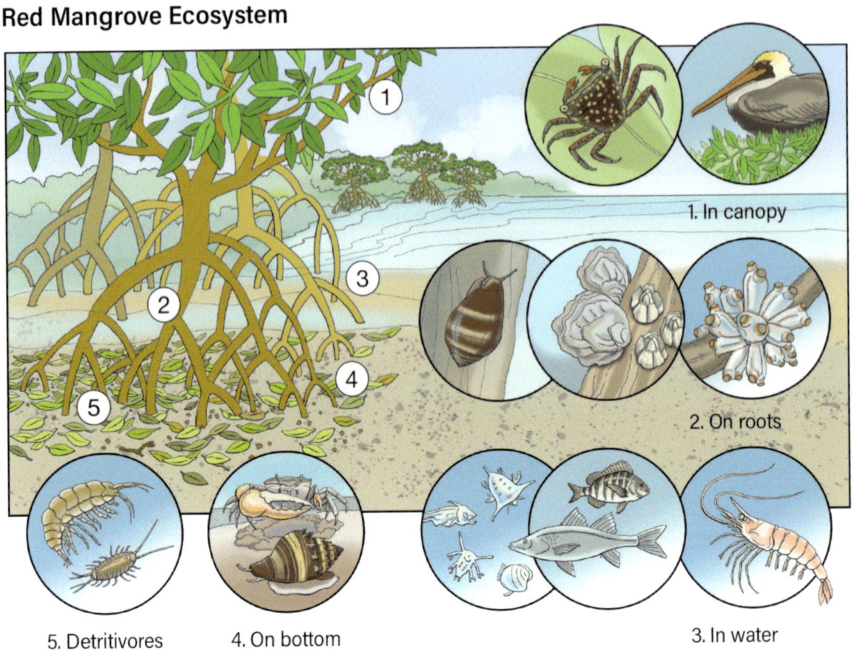


**FIGURE 1.1** A red mangrove forest (above and below the water) with some of the wide variety of organisms found in this ecosystem. *(1)* Birds, mammals, and crustaceans like crabs are found above the water and in the canopy. *(2)* Both mobile and sessile organisms such as snails, oysters, barnacles, and tunicates are found on the stilt roots between the tidal water levels. *(3)* Organisms in the water include microscopic phytoplankton (plants) and zooplankton (animals), shrimp, fish, and other free-swimming animals. *(4)* and *(5)* The benthic, or bottom, feeders include both predatory organisms and detritivores. Detritus is dead plant and animal matter that has settled on the bottom. *Illustration by Marjorie Shropshire.*

A coastal example of a working ecosystem in Florida is associated with red mangroves (*Rhizophora mangle*). These are tropical and subtropical trees or shrubs that have adapted to saline environments and are native to most of Florida, but they occur more often in South Florida's coastal zones where they sometimes form large forests at the water's edge. (For more details on red, black, and white mangroves, see below.) One adaptation red mangroves have for this environment is their stilt or prop roots that hold most of the plant above the water.

In the upper parts of the red mangrove trees, shorebirds, especially herons and egrets, make their nests. Small mammals such as raccoons or opossums also find habitat above the prop roots, where they prey on snakes, crabs, shrimp, bivalves (such as clams and mussels) and the many other creatures in this active ecosystem. Under the water, the prop roots offer shelter for larvae and juveniles of a wide variety of fish and other aquatic species, protecting them from heavy wave action and from predation when they are small. Some estimates indicate that more than one-third of marine fish species use mangroves as nurseries.[2]

The importance of mangroves is widely recognized and there are laws that regulate their management in Florida, which is covered later in this book. Mangroves are often used as part of living shoreline plantings to add structure and resilience.

All of Florida, and indeed the whole surface of the Earth, is a series of interconnected ecosystems.

## Natural Shorelines

Waves, currents, and tides shape shorelines. Wind is another major factor once the sand is above sea level. Many coastal landforms can be attributed to moving sand via longshore drift and long-term rising or falling sea levels. There are both high-energy and low-energy coastal areas in Florida, depending on wind direction, fetch (the distance the wind blows unobstructed across open water), and the profile of the bottom. High-energy coasts are usually adjacent to the Atlantic Ocean or the Gulf of Mexico and have strong wave action that suspends and exports fine particles and leaves behind only larger, coarser particles, forming sandy beaches. Such high-energy shorelines rely on a constant supply of new sand from connected areas to remain stable. Low-energy coasts are protected by land or plants in estuaries or bays so that they are subject to less natural wave energy, resulting in more accumulation of finer sediment particles; this allows, too, for the persistence of organisms such as seagrass, salt marsh plants, oysters, and mangrove trees that cannot survive the shifting sands and crashing waves of open beaches. In areas frequented by boaters, boat wakes can add artificial waves to otherwise low-energy systems and can be surprisingly strong drivers of erosion.[3]

## Florida's Freshwater Ecosystems

While most freshwater ecosystems are not in coastal areas, they play an important role in coastal ecosystems because they supply the flow of freshwater to the coastal areas both at the surface, via rivers and streams, and from below, via the aquifers, springs, and groundwater exchange.

## AQUIFERS

Aquifers are underground networks of porous rocks that hold water and allow water to move through the holes within the rock. When aquifers intersect with the surface of the soil, they are the source for springs. The Floridan aquifer is found beneath the entire state of Florida and is divided into upper and lower aquifers, separated by layers of rock or clay. There are several smaller aquifers nearer to the surface in various parts of the state. Aquifers are recharged when precipitation percolates into the soil.

## SPRINGS

There are more than seven hundred springs in Florida, which is the largest collection of natural springs on the planet! Thirty-three are first-magnitude springs with outputs of more than 750 gallons (2,839.1 liters) per second. Eight billion gallons (30.3 billion liters) of mostly 72°F (22.2°C) water flow from Florida's springs each day, and all that water ends up in coastal areas. Most springs are located in western Central and North Florida.

## LAKES AND PONDS

There are thousands of lakes and ponds throughout Florida. They provide important habitat for a wide variety of both aquatic and terrestrial wildlife. Moisture-loving plants grow at the edges of lakes, and plants growing in the water can be emergent (rooted in the bottom but growing upright above the surface), submerged, or floating. Many spring-fed lakes serve as headwaters for rivers, while other, mostly low-lying lakes may be an endpoint for streams and rivers; these lakes, with no outflow, interface with the local groundwater that may percolate into an aquifer.

Florida's largest lake is Lake Okeechobee, which covers 730 square miles (1,890.7 $km^2$) with an average depth of only 9 feet (2.7 m). The main tributary is the Kissimmee River to its north, and before the flood-control dike was built around the whole lake, it overflowed its southern bank into the Everglades in an area known as the "River of Grass" that was miles wide and only one or two feet deep. Marjory Stoneman Douglas's 1947 book, *The Everglades: River of Grass,* first publicized human-caused problems in this important ecosystem.

Because of Florida's typical wet and dry seasons, a common pond type found across much of the state is the shallow seasonal pond—typically 2 to 5 feet (0.6–1.5 m) deep and often 100 or more feet (30.5 m) across. Standing water recedes in the dry winter months, leaving a muddy pond floor that continues to provide moisture required by many amphibians, reptiles, birds, small mammals, and even small fish.

### RIVERS AND STREAMS

Rivers serve as the veins forming an interconnected web of Florida's water-driven environment. Several large rivers in North Florida flow in from other states before emptying into the Gulf of Mexico. The St. Johns River is the longest river entirely within Florida, flowing northward 310 miles (498.9 km) to Jacksonville, where it takes a sharp eastern turn to empty into the Atlantic Ocean. Along the way, the river drops only 30 feet (9.1 m) in elevation, so its flow, like most other Florida rivers, is quite slow. The slow flow combined with the incoming tidal flows at the mouth of the river creates a brackish estuary for almost a hundred 100 miles (160.9 km) up the river.

## Tidal and Littoral Zones

Tides are the rising and lowering of sea level twice a day caused by the sun and moon's gravitational effects on the oceans. At both the full moon and new moon, their gravitational forces work together to create both higher- and lower-than-average tides, known as **spring tides**, which occur about twice a month. At half-moon phases, the sun's and moon's gravitational forces are perpendicular to each other, which produce **neap tides** that have less amplitude—that is, not as high and not as low. Several times a year, there may also be **king tides**, which happen when a spring tide occurs when the moon's orbit is closer to Earth. In addition, storm surges may also bring higher water levels to Florida's coastal areas.

The average tidal amplitude in Florida is 2 to 4 feet (0.6–1.2 m). The average high tide level on a shoreline is called the mean high water (MHW) line, the average low tide level on a shoreline is called the mean low water (MLW) line, and the shoreline in between these levels is the intertidal zone. The averages of king and spring tides, with levels that are both higher and lower than the mean levels, are called the mean higher-high water (MHHW) line and the mean lower-low water (MLLW) line, respectively. Coasts with greater tidal ranges are usually more buffered against storm surges than those with less tidal amplitude.[4]

The littoral zone in an aquatic ecosystem (river, lake, sea) is the transition area between the land and the body of water. Littoral zones, especially in freshwater systems, are characterized by the presence of sunlight all the way to the bottom. For marine environments subjected to tidal flows, the intertidal area is often synonymous with the littoral zone. Other descriptions of marine littoral areas include the area all the way to the edge of the continental shelf, and in Florida, depending upon the location; the continental shelf may extend for many miles. On the east coast, off St. Augustine in North Florida, the marine littoral area is more than 60 miles wide (96.6 km); off Palm Beach in South Florida, it is only

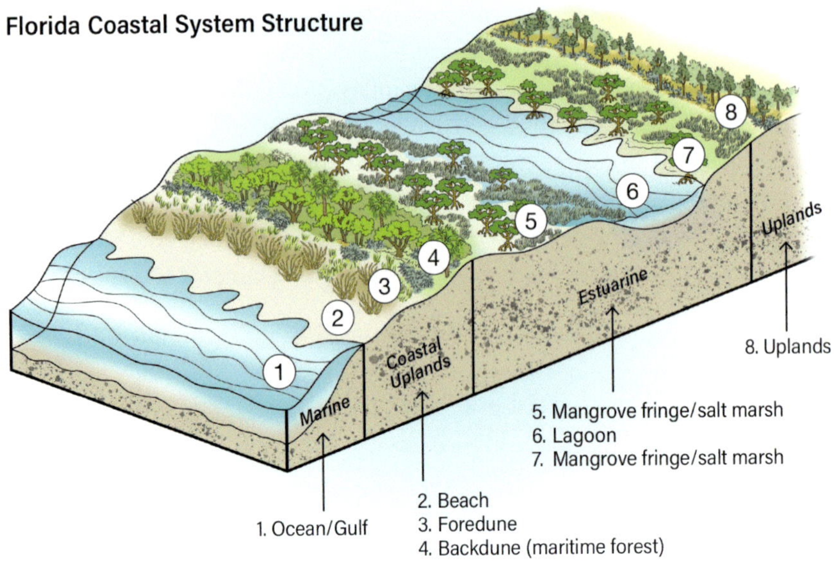


**FIGURE 1.2** Florida's coastal system structure. This diagram shows *(1)* the transition moving inland from deep water marine areas to habitats that exist at higher elevations. *(2–4)* Coastal uplands include beach and dune habitats on a barrier island. *(5–7)* Estuarine habitats, including a lagoon, are fronted on both sides by mangrove fringe and/or salt marsh, depending upon location. *(8)* Uplands inhabit the area above the mean higher-high water (MHHW) line on the far side of the estuary. *Illustration by Marjorie Shropshire.*

about one mile wide (1.6 km). The continental shelf off the west coast of Florida extends for more than a 100 miles (160.9 km).

Because light reaches the bottom and because of the regular movement of water along the shore areas, there is often a wide variety of partially submerged (emergent) to fully submerged aquatic plants in littoral zones. These plants form the basis for an ecosystem which supports both terrestrial and aquatic animals.

## ESTUARIES

Estuaries are bodies of water, often partially enclosed, where freshwater from rivers, streams, and springs mixes with the salty, tidal waters from the Atlantic Ocean or the Gulf of Mexico. Estuaries are either bays that are mostly surrounded by land or the mouths of rivers that are affected by tides; and because Florida's rivers move so slowly, the tidal prism often flows upriver for long distances. The saltwater is heavier than the fresh water, so the incoming tidal waters sink below the outflowing freshwater from the rivers and streams, which affects the dynamics of the eventual mixing of salty and fresh waters as the estuary becomes brackish.

Since most estuarine areas are protected to some degree from the full force of ocean waves, winds, and storms, these coastal areas are generally lower-energy ecosystems than the coastal areas that are exposed to the full brunt of the waves and storms from the Atlantic or the Gulf. Estuaries support many different ecosystems, including shallow open waters, salt marshes, swamps, sandy beaches, mud and sand flats, rocky shores, oyster reefs, mangrove forests, river deltas, tidal pools, and seagrass meadows.

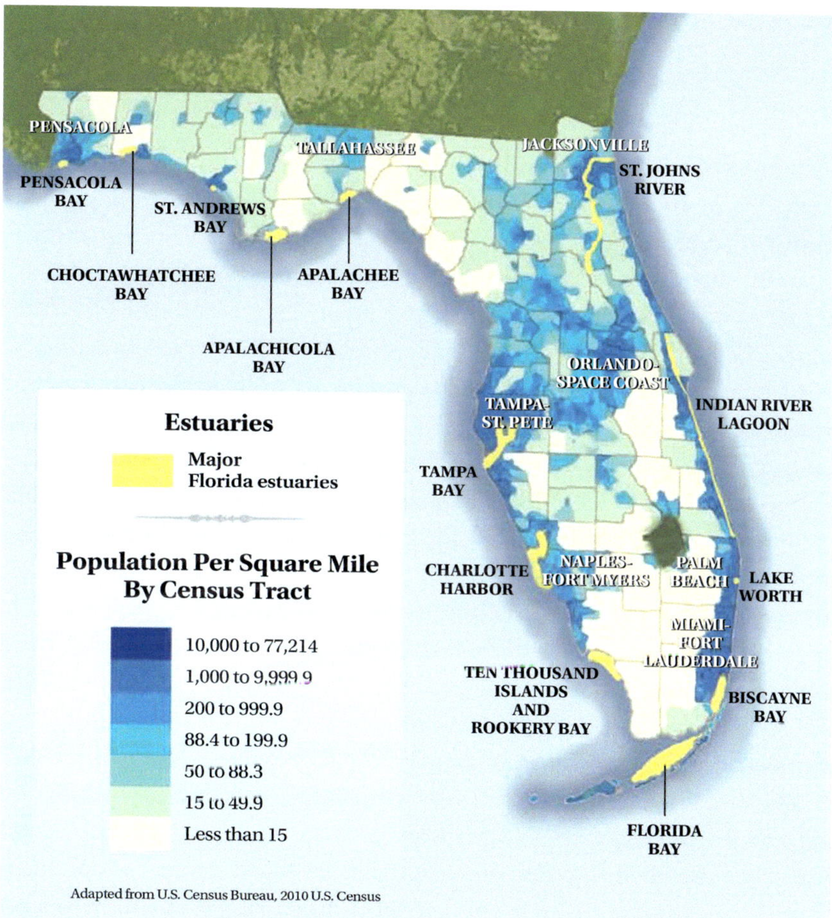


**FIGURE 1.3** Florida has thirteen major estuaries located along its coastal areas. This graphic shows how the population density varies across the watersheds that feed into each estuary (shown in yellow). *Reproduced with permission from Karl Havens, Gary Lytton, and William Seaman, A Practical Guide to Estuary-friendly Living, SGEB-64 (Florida Sea Grant College Program, 2012).*

Because these are sheltered and diverse habitats, estuarine environments are among the most productive on earth, creating more organic matter each year than comparably sized areas of forest, grassland, or agricultural land. The high productivity in the sheltered waters of estuaries provides habitat, protected spawning places, and support for so many species of birds, fish, and other wildlife; for this reason they are often called the "nurseries of the sea."

According to Florida Sea Grant, there are thirteen major estuaries in Florida. Starting from the northeastern coast, they are as follows: the St. Johns River, the Indian River Lagoon, Lake Worth, Biscayne Bay, Florida Bay, the Ten Thousand Islands and Rookery Bay, Charlotte Harbor, Tampa Bay, Apalachee Bay, Apalachicola Bay, St. Andrews Bay, Choctawhatchee Bay, and Pensacola Bay.

## Dunes and Maritime Forests

More than 60 percent of Florida's coastline (not including the Florida Keys) is sandy. The source of and composition of Florida's sand varies depending upon location. In North Florida, both on the Atlantic coast and the Gulf coast, the sand is mostly quartz granules that have been carried from the eroding Appalachian Mountains over the ages by long rivers that empty into the Atlantic Ocean and into the Gulf of Mexico. In South Florida, most of the sand is calcium carbonate, which is made up of pieces of shells and skeletons of reef-living marine organisms, including corals, mollusks, and some microorganisms. Some of these sand particles are created by erosion by currents and waves, but more of this sand is due to bioerosion, which occurs when fish and other marine organisms consume the coral and other shellfish and then excrete the leftover pieces.

Dunes and barrier islands are dynamic sandy habitats moved by the longshore current and drift. When the longshore current deposits sand along the coast into quieter waters, the sand and sediments accumulate, creating spits or beaches. Once land is built, and when the wind is greater than 13 miles per hour (20.9 km/h), it also carries sand. And it is this wind-borne sand that builds up and shapes the dunes.[5] Eventually, barrier islands formed along most of Florida's eastern coastline as sand was carried along with currents; and over time, these sand islands breached and moved in reaction to weather events and currents. The dunes and barrier islands generally move in the direction of the longshore drift and toward the uplands at about 1 to 2 feet (0.3 to 0.6 m) per year on average, but the rate of movement varies greatly depending upon local currents and the occurrence of hurricanes and other significant storms.

Most dune systems both on sandy capes and on broader barrier islands consist of sets of ridges and valleys behind the foredune, which is the dune closest to the water. These form over time and the older dunes are usually covered with and held in place by a variety of highly specialized plants that can survive in a dune's shifting sands and the salty environment. These tough plants use various methods to survive in this difficult environment. Such methods include being able to quickly regrow after being buried by sand and making sure their seeds are buried deeply. The upper beach and foredune areas in Florida are populated by just a few species of plants that can survive these conditions, most notably, sea oats (*Uniola paniculata*), dune sunflower (*Helianthus debilis*), and railroad vine (*Ipomoea pes-caprae*). These plants help shape the dunes because they interrupt and slow down the wind, which in turn causes more wind-borne sand to pile up around each plant. The upper beach and foredune areas may be subjected to significant reshaping by severe storm events with their high winds and surges of high water. After such storms, the plants must start over, recolonizing the bare sand.

The plant diversity increases somewhat through the transitional zone on and between the secondary dunes behind the foredune. This zone continues into the upland plant community or ecosystem. The transition zone plant communities vary greatly depending upon the local climate and the type of sand, but in most locations the taller woody plants are "pruned" by the salty winds and have a wind-swept shape.

**FIGURE 1.4** Beach dune in northeastern Florida. Beach and foredune areas may be reshaped by severe storm events, and after these storms, the tough dune plants recolonize the bare sand. *Photo by Ginny Stibolt.*

## HOW PLANTS WORK

### PHOTOSYNTHESIS AND RESPIRATION

Green plants and other organisms with chlorophyll create the basis for all terrestrial and most of aquatic ecosystems through **photosynthesis**, which is the process by which carbon dioxide (from the air) and water (from the soil or air) with the energy from the sun combine to produce sugars with oxygen gas given off as a byproduct. These carbon-based sugars provide food for all terrestrial life by which plants, animals, fungi, bacteria, and other organisms take in oxygen and sugars, one way or another, to gain energy for life through **respiration** and during this process they release carbon dioxide and water. Photosynthesis and respiration are equal and opposite reactions:

**Photosynthesis:** $6CO_2 + 6H_2O + \text{sun's energy} \rightarrow C_6H_{12}O_6 + 6O_2$

**Respiration:** $C_6H_{12}O_6 + 6O_2 \rightarrow \text{energy for life} + 6CO_2 + 6H_2O$

The other item to note here is that photosynthesis removes carbon dioxide, our most abundant greenhouse gas, from the air. While carbon dioxide is released back into the air during respiration, plants not digested but stored in peat or sedimentary rock formations sequester that carbon and those formations become carbon sinks. (See "Blue Carbon, Peat, and Peatlands" sidebar, page 21.)

## Mangrove Forests

According to NOAA, "Mangroves are a group of trees and shrubs that live in the coastal intertidal zone."[6]

The term *mangrove* can describe either a single plant, a single species, or the whole plant community that grows along low-energy saline or brackish shorelines. Mangrove trees and shrubs are facultative halophytes, so they may not require salty environments for good growth, but they tend to grow in salty environments because of reduced competition from other plants. Salty environments are harsh for plants, and the mangroves use various energy-consuming processes, including extensive filtration and the exudation of salt through special pores to filter out the salt and take in enough usable freshwater for growth. Most mangroves have developed various methods to conserve their freshwater (like desert plants), such as extensive root systems, succulence, thick cuticles, and relatively low rates of water evaporation through pores on the leaves or bark. They may also absorb freshwater from rain through their bark, especially through the surfaces of their drop roots, which are roots that arise from the branches and are usually above mean high tide.

**FIGURE 1.5** Red mangroves (*Rhizophora mangle*) are easy to identify because of their stilt or prop roots extending out into the water. *UF/IFAS file photo by Tyler Jones.*

Depending upon the reporting authority, there are between 30 and 80 mangrove species worldwide; these species belonging to 16 to 20 different families. Mangroves are not necessarily related botanically, but they are grouped by their ability to grow at the edge of the sea. Most mangroves are viviparous, which means their seeds germinate while still attached to the parent tree. This allows seedlings to drift away with the tides and then quickly establish themselves—because energy is already available from the photosynthesizing seedling in the propagule. (See sidebar "Viviparous Germination.")

Mangroves are woody plants, but since they typically grow in environments that have a constant water supply and consistent temperatures, their annual rings are usually indistinct and result from wet and dry seasons affecting water and soil salinity. Mature mangroves grow slowly, so their growth rings are small, and people analyzing them usually need magnification to determine their age.[7] Most mangroves do not spread vegetatively via roots or shoots. This means that restoration of mangrove forests uses propagules for each new plant, not cuttings from branches or divisions from mature plants. This may be a limiting factor, but it also ensures better genetic diversity.

Mangroves growing in thickets along tropical and subtropical shorelines absorb nutrient outflows from the land, absorb the wave action from open waters, and build new soil as they slow down and hold onto passing sediment and floating

## VIVIPAROUS GERMINATION

Viviparous means "born alive"; humans and many other animals are viviparous.

Most seed-producing plants (both angiosperms and gymnosperms) develop seeds after sexual fertilization, which are then dropped or separated from the plant one way or another in a dormant state. When properly stimulated, usually with moisture and contact with soil, the seeds germinate and begin to grow into a new plant.

A viviparous plant is one in which the seed germinates while still attached to the parent plant and, once germinated, the remaining fruit and the seedling may already have roots, a stem, and leaves. For mangroves, these seedling structures, released from the parent plant, are known as propagules. When dropped, they can float away until they land on a favorable site. The leaves and green stems of the propagules, already capable of photosynthesis at the point of landing, can quickly supply the energy necessary to fuel root penetration. A mangrove propagule may float for many months and not experience any ill effects; indeed, some species have a minimum amount of time to float around before they settle, which may be an adaptation that ensures the offspring are at some distance from the parent plant.

A semi-viviparous or crypto-viviparous plant is one in which the seed germinates within the fruit but does not break through the fruit wall until after it is released from the parent plant—either as it is floating or when it lands in an appropriate spot. Both the black mangrove and the white mangrove fall into this category.

**Mangrove Propagules**

**FIGURE 1.6** After mangrove propagules are released from the parent plant, they float until they land on a favorable site, where they can take root. *(1)* Red mangrove (*Rhizophora mangle*) propagules are the largest at 6–10 inches (15–25 cm) long. *(2)* Black mangrove (*Avicennia germinans*) propagules are about the size of a lima bean. *(3)* White mangrove (*Laguncularia racemosa*) propagules are smaller, and vase shaped.

*Illustration by Marjorie Shropshire.*

debris carried in the tidal waters. This improves the local water quality for other marine ecosystems in which silt and other debris carried in the water can cause problems for many organisms. This is especially important for coral reefs and seagrass meadows. Mangroves also create habitat for many types of wildlife, including reptiles, amphibians, birds, fish, and a wide assortment of invertebrates, including crustaceans.

When they die, the plants and animals that live in the mangrove forest ecosystem add to the mix of floating sediment and debris. All this detritus buildup forms peat soils in an oxygen-poor environment. This is why some mangrove species, such as red mangrove, have adapted to this environment by developing stilt or prop roots that extend from the lower parts of the trunk or drop roots that extend from branches or other upper parts of the plant. The aerial portions of the roots contain lenticels, which are special pores that keep water out during high tide but allow oxygen to enter the plant when they are above water. Other species, such as the black mangrove, develop extensive shallow roots that radiate out from the main trunk with vertical root extensions called pneumatophores, which also contain lenticels for gas exchange. The height of the pneumatophores varies depending on local conditions. The white mangrove develops lenticels on the lower trunk, and it occasionally develops a few pneumatophores when local tides are irregular. These special root structures that allow mangroves to grow in oxygen-poor soil are also vulnerable to clogging by oil pollution and fine sediments; the mangroves may drown during periods of prolonged flooding.[8]

The peat formed under mangroves can be held in place for thousands of years, and this means that the mangrove forests sequester carbon at a far greater rate than other forests, including rainforests.[9] Because these carbon-sequestering soils are below the surface of the water, they are often referred to as "blue carbon." (See sidebar "Blue Carbon, Peat, and Peatlands.")

According to the Florida Department of Environmental Protection, there are about 6,000,000 acres (24,281 km$^2$) of mangrove forests in Florida, 90 percent of which are located in the four southernmost counties of Florida. Mangroves spread on new land formed at the seaward edge of these forests and into saltmarsh areas, especially in more northerly ranges. Many ecologists consider this to be a natural succession and that mangrove forests are the climax ecosystems for tidal areas.[10]

Three mangrove species are native to Florida: red, black, and white; and because it so often grows in the same habitat, another native, green buttonwood, is often thought of as Florida's unofficial mangrove or a mangrove associate. When these mangrove species, along with the buttonwood, grow along the same shoreline, each grows in a different area. This zonation is determined by tidal flows, elevation of the land, and the salinity of the soil and water. Mangroves are tropical and subtropical, and so have occurred mostly in South Florida and Central Florida. In the last few decades, however, the mangroves have been moving northward and often establish themselves on the waterside edges of established salt marshes.[11]

**Red mangrove (*Rhizophora mangle*)** is in the Rhizophoraceae family. When Florida's three mangrove species grow together, this species usually grows farthest out in the water. Tangles of prop or stilt roots emerge from their trunks, grow into the bottom, and act as stilts to increase the plant's stability. Since the prop roots disturb and slow down the water flow, they facilitate the capturing of sediments from the water. The red mangroves are sometimes called walking trees because of these specialized roots. They also have drop roots, which emerge from branches and hang down above the water. The red mangrove propagules are 6 to 10 inches (15 to 25cm) long and look like an elongated seed pod; but each propagule is a plant capable of rooting quickly when it lands in a suitable location. Red mangroves are found in most of Florida's coastal areas and are also native to the Caribbean Islands, Mexico, Central America, South America, and tropical western coastal areas in Africa. These are the mangroves most often used for living shorelines in Florida.

**Black mangrove (*Avicennia germinans*)** is in the Acanthaceae family. Black mangroves tend to grow in waters shallower than the waters red mangroves grow in and deeper than the waters white mangroves grow in. The black mangrove produces pneumatophores, straw-like extensions from its roots, to supply needed oxygen and gas exchange to the roots. Its propagules contain a fruit with a germinated seed inside and are about the size of a lima bean; but when they land, four layers split open, and the fuzzy root emerges first. Black mangroves are found in most of Florida's coastal areas, and in Georgia and along Texas's shorelines; and they are also native to the Caribbean Islands, Mexico, Central America, South America, and tropical western coastal areas in Africa. The black mangrove's range is a bit more extensive and contiguous than the red mangrove's, potentially because it is the most cold-tolerant mangrove species in the Atlantic/Caribbean region.

**White mangrove (*Laguncularia racemosa*)** is in the Combretaceae family. This species is less cold-tolerant and has not migrated as far north as the black and red mangroves. The white mangrove usually grows on the landward side of the other two species and sometimes has root adaptations like those of the black mangrove, but such adaptations are less extensive. Like the other mangrove species, it handles the salty environment, but one of its adaptations is that it exudes salt through its leaf pores, so the leaves end up looking white. Its propagules are small vase-shaped fruits with sprouted seeds inside, which are ready to grow and root quickly when they land. White mangroves are found in Central and South Florida's coastal areas

## BLUE CARBON, PEAT, AND PEATLANDS

*Blue carbon* is the term used to describe the removal of carbon dioxide from the atmosphere by oceanic and coastal ecosystems. The salt marsh and coastal wetland plants involved in this carbon sequestration include various types of algae, phytoplankton, seagrasses, and mangroves. The carbon these plants remove from the atmosphere is sequestered below sea level, even in coastal ecosystems. Globally, these plants remove 83 percent of the carbon and produce up to 80 percent of the oxygen in the atmosphere. Coastal habitats cover less than 2 percent of the total ocean area, but they account for approximately half of the total carbon sequestered in sediments. Seagrass meadows, salt marshes, and mangrove forests are sometimes referred to as "blue forests" in contrast to upland-based "green forests."

In a normal, terrestrial, aerated soil ecosystem, carbon-rich organic materials (dead plants, animals, fungi, and their waste) decompose relatively quickly due to the activity of soil organisms, including microbes (bacteria, fungi, and others), worms (ranging from microscopic nematodes to large earthworms), insect larvae, and other critters that all derive energy for life from the soil's organic material. By contrast, peat is only partially decomposed organic matter formed in anaerobic environments with little or no oxygen—a condition called hypoxia. This lack of oxygen means that most of the soil organisms, which would normally digest the organic matter, are not present, so the organic materials continue to accumulate.

Peat is formed in several types of ecosystems, including peat bogs, pine barrens, some sections of rainforests, mangrove forests, salt marshes, and seagrass meadows. In most of the habitats where peat is formed, it builds up over the long term—in some cases over thousands of years. Worldwide, peatlands make up about 3 percent of the land but account for over 40 percent of the soil carbon sequestration.[a]

Peat is formed under salt marshes, mangrove forests, and seagrass meadows because there is little or no oxygen in the soil due to the tidal flooding and draining cycle of saline or brackish water. These plant communities form the basis for working ecosystems with populations of algae, diatoms, phytoplankton, birds, terrestrial animals, and aquatic animals, which add their waste to the detritus of leaves and other plant and animal parts. In addition, the sediments carried in by the tides and the currents are dropped as the water movement is slowed by the presence of vegetation. The peat formed in salt marshes and mangrove forests accumulates at a much faster rate than peat formed in bogs because the plants are larger, there is a presence of a community of animals, and most significantly because of the addition of sediments carried by water with the tidal flows and other currents. Studies have shown that in many cases, peat formed by salt marshes and mangrove forests accumulates at rates up to 5 inches (12.7 cm) per year.[b]

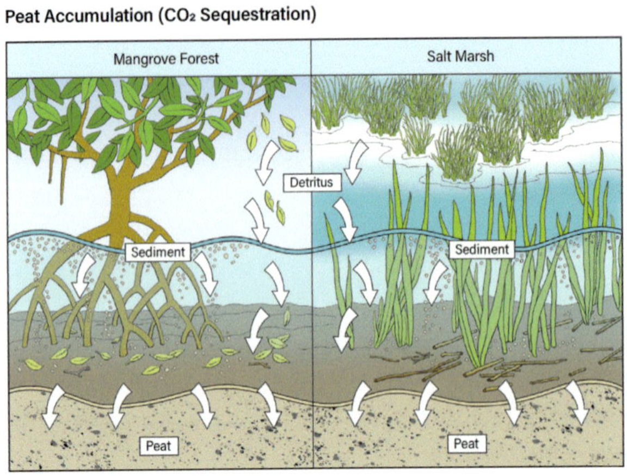


**FIGURE 1.7** In mangrove forests and salt marshes, sediment and detritus are captured by roots and rhizomes, where it accumulates; and since there is little or no oxygen in the soil under the water, peat forms and may be held in place for thousands of years. Peat sequesters the carbon from the dead plant and animal parts and plays a role in the "blue carbon sequestration." *Illustration by Marjorie Shropshire.*

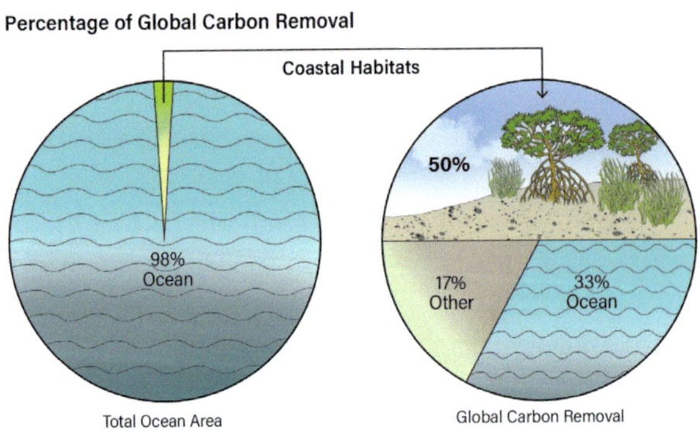


**FIGURE 1.8** Ocean and coastal ecosystems remove 83 percent of global carbon; of that, 50 percent is accomplished by coastal ecosystems, even though they are only 2 percent of the total ocean area. Data from The Blue Carbon Initiative, https://www.thebluecarboninitiative.org/. *Illustration by Marjorie Shropshire.*

[a] UN Environment Programme,"Peatlands Store Twice as Much Carbon as All the World's Forests," accessed January 31, 2025, https://www.unep.org/news-and-stories/story/peatlands-store-twice-much-carbon-all-worlds-forests.

[b] Anna McIvor, Tom Spencer, Iris Möller, and Mark Spalding, The Response of Mangrove Soil Surface Elevation to Sea Level Rise, Report No. 3: Natural Coastal Protection Series (The Nature Conservancy and Wetlands International, 2013), https://www.conservationgateway.org/ConservationPractices/Marine/crr/library/Documents/mangrove-surface-elevation-and-sea-level-rise.pdf.

and are also native to Texas, the Caribbean Islands, Mexico, Central America, South America, and tropical western coastal areas in Africa.

**Buttonwood (*Conocarpus erectus*)** is also in the Combretaceae family. A small mangrove-like tree that naturally occurs in coastal areas, but it is not viviparous. It grows best in full sun or partial shade at the edges of fresh, brackish, or saltwater marshes. It is useful for building upland buffers in Central and South Florida (see chapter 8), but it can also withstand the rigors of urban settings and makes a durable street or parking-lot tree. They are found in Central and South Florida's coastal areas and are also native to Texas, the Caribbean Islands, Mexico, Central America, South America, and tropical western coastal areas in Africa.

Mangroves, especially the red mangrove, have been successfully used in building living shorelines. They have a high survival rate and their ability to protect the shoreline, hold the soil, and provide habitat are important features. (See chapter 3 for details.)

## Salt Marshes

Salt marshes form in many low-energy, tidal shoreline areas in Florida, and while estimates vary, as of this writing, most agree that there are more than 420,000 acres (1,699.7 km$^2$) of salt marshes in Florida. About 20 percent of the salt marshes occur as fringes on the upland side of mangrove forests. Like mangrove forests, salt marshes are populated by only a few species of plants that can live in salty or brackish environments with little or no oxygen in the underlying peat soils. They are characterized by vast expanses of grasses, rushes, and sedges.[12]

Despite these harsh conditions, the net primary production (photosynthesis minus respiration of the organisms that live there) in salt marshes is among the highest of the world's ecosystems, and this high production forms the basis of both terrestrial and marine food webs. Only about 10 percent of the biomass of the salt marsh plants is consumed while alive. After plants die, the detritus may be consumed to some degree, but most of it becomes part of the underlying peat soils.

Like mangrove forests, salt marsh ecosystems provide important habitat for a wide variety of both aquatic and terrestrial wildlife, capture sediments from the water, intercept pollutants and nutrients from upland runoff, and protect the uplands behind them from wave action and storm surges. Because of the tidal flows and freshwater outflows from upland areas, salt marshes are riddled with outflow creek indentations that accommodate the twice-daily drainage that occurs during low tides. These areas provide critical habitat for many fish and other aquatic species.

**FIGURE 1.9** Two views of salt marsh ecosystems. *(A)* The Great Blue Heron (*Ardea herodias*) is just one of the top predators in salt marshes. This productive ecosystem provides habitat for a wide variety of animals, including many of our commercially important fish. *(B)* Salt marshes are in the intertidal zone, between mean low water (MLW) and mean high water (MHW) and covered by tidal water twice a day. The natural drainage rivulets meander dramatically, as seen in this overhead image. *UF/IFAS file photos by Tyler Jones.*

Salt marshes may look uniformly flat, but berms and ridges build up over time, and even just a slight elevation change may mean different tidal flow and upland output; so these ridges are often inhabited by a different set of plants. This zonation of salt marshes is a product of the nature of neighboring upland ecosystems and adjacent bodies of water. Some parts of salt marshes may be above mean high tide and are flooded only twice a month during spring tides and even less often during king tides or storm surges. This area is known as high marsh, and it supports a higher diversity of plants; but these plants still need to be salt tolerant. (See page 11 for more details on tides.)

Florida's dominant salt marsh species include the following: black needle rush (*Juncus roemerianus*), which occurs along higher marsh areas; saltmeadow cordgrass (*Spartina patens*), which grows in high marsh areas that are only periodically inundated;

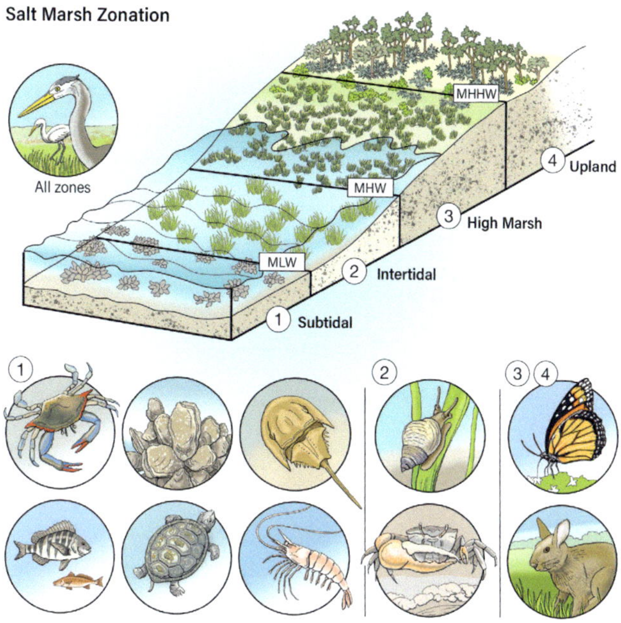


**FIGURE 1.10** Salt marsh zonation and associated creatures. *(1)* The subtidal zone is below the mean low water (MLW) line and will be covered by water most of the time. *(2)* The intertidal zone is between the MLW and the mean high water (MHW) lines and will be covered and uncovered by water once or twice a day. The intertidal zone includes the low marsh. *(3)* The high marsh zone is generally flooded only during the most extreme (king) tides and larger storm events up to the mean higher-high water (MHHW) line. *(4)* The upland areas are rarely flooded but may be exposed to salt spray regularly. *Illustration by Marjorie Shropshire.*

smooth cordgrass (*Spartina alterniflora*), which is found in the lowest areas that are most frequently inundated; and sawgrass (*Cladium jamaicense*), which is actually a freshwater sedge that sometimes grows along the upper edges of salt marshes. All are tolerant of the salt in sea spray. (Some botanists place the *Spartina* species in the *Sporobolus* genus, so saltmeadow cordgrass could be referred to as [*Sporobolus pumilus*], and smooth cordgrass could be referred to as [*Sporobolus alterniflorus*].)

Salt marsh vegetation physically dampens wave action by increasing friction, and this helps marshes resist erosion. Plus, the dampening of the waves causes sedimentation deposits and clearing of the water. Their effectiveness is surprising. Even narrow bands of marsh (about 100 feet [30.5 m] wide) have been shown to be effective at attenuating most of the incoming wave energy, and waves were eliminated by around eighty meters of marsh.[13]

Salt marsh systems are naturally dynamic and constantly changing and thus provide excellent habitat and protection for the upland areas behind them. Living shorelines are often designed to simulate these and other functions of salt marsh ecosystems.

## Oyster Reefs

Eastern oysters (*Crassostrea virginica*) have a natural range that extends from Canada to Mexico, including all of Florida's brackish, lower-energy coastal waters. Oysters filter feed, consuming plankton and other microscopic aquatic organisms. As they filter the water to feed, they also absorb nutrients, sediments, heavy metals, and toxins from the water. A single oyster can filter more than fifty gallons of water per day, depending upon the oyster's location, size, and water conditions, so a whole collection of oysters in a reef can significantly improve water quality and clarity.[14] This can sometimes result in clearer water, allowing greater light penetration, which promotes better growth of submerged aquatic vegetation, especially seagrasses.

Oyster larvae, known as spat, settle on hard surfaces (oyster shells preferentially). Once they are attached, they remain there for the rest of their lives. The reefs formed as more and more oysters grow on and around each other are filled with nooks and crannies, which are essential habitats for many other species of fish, shellfish, crustaceans, and more. The reefs occur naturally either in subtidal areas, where they are usually covered by water, or in intertidal areas, where they are exposed to the air during low tide. Oysters are tolerant of drying out and they can endure lower salinities and even short periods of freshwater—under such conditions, they simply close their shells until the environment becomes favorable for feeding and filtering again. Logically, those oysters that are underwater for optimal periods of time and are not frequently flushed with too much freshwater will have ample opportunity to feed by filtering the water and will generally grow at a faster rate than those that are regularly stressed.

Oysters are considered a foundation species for many of Florida's coastal brackish-water ecosystems because they form reefs that play several critical roles in the ecosystem: filtering nutrients and particles from the water column, reducing wave energy that would otherwise contribute to shoreline erosion, and providing habitat and food for numerous fish species, marine invertebrates, shore birds, and more.

Many organisms depend on oyster reefs for forage and refuge, contributing to biodiversity in the estuary. As many as 300 species have been documented as associated with eastern oyster reefs.[15] By virtue of forming physical habitat as reefs and influencing the environment around them through filter feeding, oysters are also considered ecosystem engineers. The complex and intricate reef structure created by oyster shells also influences the physical environment by altering current flow and patterns, which can in turn influence the suspension and delivery of phytoplankton, or microalgae, just above the reef.[16] This can enhance food delivery to the oysters, which filter and consume the phytoplankton from the water column, as well as to other organisms with and attached to the oysters, such as mussels, sea squirts, and sponges. This alteration of flow patterns can also influence the deposition of the oysters' organic-rich biodeposits, which in turn support organisms such as worms, barnacles, crabs, and more. The microbial fauna found within and around the oyster reef often facilitates the cycling of nutrients that can remove excess nitrogen from the overlying waters.

Since oysters are filter feeders, particles, such as microalgae, are sent into the oysters' mouths where they are digested, and waste is excreted as feces. Inorganic particles, such as sediment, are usually rejected, bound together, and deposited as "pseudofeces." They are not digested. Biodeposits include both feces and pseudofeces.

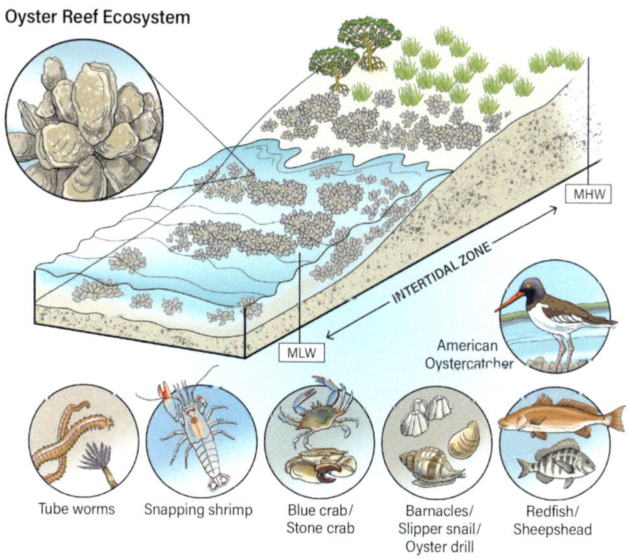


**FIGURE 1.11** The complex surface area and structure of an oyster reef physically influences the environment by dampening waves. This slowing of water causes sediments to drop in the area. The reef's complex surface also provides habitat for a wide variety of creatures. *Illustration by Marjorie Shropshire.*

Oyster reefs also stabilize sand bars and the shorelines behind them by providing protection from wave energy. The three-dimensional habitat of cemented oyster shells provides refuge for a variety of species and can act as "living breakwaters" while forming significant reefs. Intertidal reefs can play an important role in stabilizing and protecting adjacent habitats and shorelines. A healthy, self-sustaining reef can continue to provide protection from waves as successive generations of newly settled oysters continue adding shell and structure to this natural form of coastal defense. By contrast, oyster reefs that experience extreme or continuing stress events, such as freshwater discharges, physical destruction, or chronically impaired water quality can lose their ability to provide coastal protection. Dead oyster shells will get broken up and washed away by wave energy, whether natural or human-caused (boat wakes). Oyster reefs that are subject to constant wave energy without repopulation by new oysters eventually deteriorate.

On a larger scale, oyster reefs can also help form estuaries themselves. In the Big Bend region of Florida, oyster reefs formed offshore in the Gulf of Mexico and impounded water from rivers entering the Gulf, creating a basin where

**FIGURE 1.12** This oyster reef in Florida's Big Bend area stabilizes the shoreline behind it. Healthy oyster reefs are self-sustaining since they attract new spat, which continue to build reef structures. Healthy reefs help prevent erosion and reduce wave energy, which causes sediment to be deposited in the area. *UF/IFAS file photo by Tyler Jones.*

freshwater and saltwater could mix. This type of bar-built estuary was created by oyster reefs formed long ago.[17]

Building oyster reefs is commonly used for living shorelines in areas with existing oyster populations that can inhabit the new reefs, so they can develop further. (See chapter 3 for details.)

## Seagrass Meadows

Seagrasses are a collection of about seventy-five species of vascular flowering plants (monocots) that have adapted, over millions of years, from living on land to life underwater. There are five species that are native to Florida. The three that are the most common are shoal grass (*Halodule wrightii*), manatee grass (*Syringodium filiforme*) and turtle grass (*Thalassia testudinum*), and these species can form continuous meadows. The other two native species, star grass (*Halophila engelmannii*) and paddle grass (*Halophila decipiens*), and two non-native species of seagrass in the same genus (*Halophila ovalis* and *Halophila stipulacea*), are diminutive species that are less common and generally do not form large meadows. Widgeon grass (*Ruppia maritima*) is not a true seagrass based on its evolutionary history, but it is a salt-tolerant native submerged vegetation species that is often found in association with seagrasses, especially in more brackish areas.

Seagrasses flower only for a short time each year and, like many land plants, flowering is linked to the seasons. Small crustaceans (amphipods and isopods) and worms may pollinate seagrass flowers, and pollination can also occur through waterborne pollen transport.[18] Since seagrasses evolved from terrestrial species, they are vascular plants with veins that transport sugars, nutrients, and oxygen throughout the plant. The veins also contain air pockets, called lacunae, that help keep the seagrass leaves floating in the water. Seagrasses are not related to seaweeds (also known as macroalgae), which are not vascular and do not flower.

Unlike land plants that have oxygen available from the soil around their roots, seagrasses growing in the waterlogged sand, peat, or mud have very little oxygen available. Many seagrasses combat the anoxic soil conditions by releasing oxygen from their root systems to create an oxygenated buffer zone in the sediments.[19] Seagrasses are also able to absorb nutrients, oxygen, and carbon dioxide from the water through their leaves as well as through their roots. The dense seagrass roots and rhizomes also serve as anchors in the sediment, and some species store nutrients in the form of sugars or starches in their roots.

Seagrass meadows help to maintain water clarity by trapping fine sediments and particles with their leaves and by stabilizing the sea bottom with their extensive root systems. Like mangrove forests and salt marshes, seagrasses also build a type of peat in anaerobic underwater soil, but to a lesser extent. The seagrasses make up most of the rooted submerged aquatic vegetation (SAV) in coastal Florida and play an important role in a balanced aquatic ecosystem.

According to Florida Fish and Wildlife Conservation Commission, Florida has more than two million acres (8,093.7 km$^2$) of seagrass meadows along its coastlines and within its estuaries. They are especially abundant in the Big Bend area of Florida, from Tarpon Springs northward to Apalachee Bay on the Gulf coast, and as far north as Volusia County on the Atlantic coast.

Seagrass meadows provide food, shelter, and essential nursery areas for more than 500 species of fish and other marine life. Some of these marine organisms can be found in seagrass throughout the year, while others remain in seagrass beds for only part of their life cycles or during certain seasons.

A wide range of marine organisms that live in or near the bottom often live within the habitat offered by seagrass meadows, where they can be protected from strong currents, hide from predators, and take advantage of the shelter provided in the dense canopy and root systems. Some of the creatures that live in or depend on the seagrass ecosystem include manatees, sea turtles, stingrays, crabs, snails, starfish, anemones, clams, worms, and countless fish species. Many of the smaller organisms are so well adapted to this ecosystem that their bodies are camouflaged to look like seagrasses, so predators have a harder time finding them.

## Coral Reefs

Florida is the only state in the continental United States with extensive shallow coral reef formations near its coasts. Florida's coral reefs stretch over 350 linear miles (563.3 km) from Dry Tortugas National Park, west of the Florida Keys, to the St. Lucie Inlet in Martin County. Roughly two-thirds of Florida's coral reef lies within Biscayne National Park and the Florida Keys National Marine Sanctuary. The reefs stretching north of Biscayne National Park and the marine sanctuary are managed by the Florida Department of Environmental Protection's Coral Reef Conservation Program with input from the Southeast Florida Coral Reef Initiative (SEFCRI).

Coral reefs are among the most species-rich ecosystems on earth, rivaled only by tropical rainforests in their diversity. Coral reef systems have an outsized influence on underwater and shoreline ecosystems: They occupy a mere 1 percent of the underwater area, but they support 25 percent of the sea life.[20]

Corals are animals, and like oysters, they are free floating when young, before they settle on a hard surface, where they then spend their adult lives. As adults, corals are generally referred to as polyps. Corals can reproduce asexually by budding. They also reproduce sexually and are hermaphroditic, with both male and female sexual organs. They reproduce by releasing their eggs and sperm into the seawater on certain full moon nights during the year.

**Stony or hard corals** (Scleractinia corals) have cylindrical bodies with a mouth and fringe of six (or multiples of six) stinging tentacles at the top. The polyps extract calcium from seawater and combine it with dissolved carbon dioxide to construct the elaborate calcium carbonate ($CaCO_3$) exoskeletons. This process is slowed down if the sea is too acidic, because $CaCO_3$ is alkaline.

Almost all hard corals contain symbiotic algae called zooxanthellae, which provide sugars and oxygen through photosynthesis. This restricts most of these corals to shallow areas with clear water, because the algae need enough light for photosynthesis. Most stony corals form colonies in which the polyps may be less than 0.5 inches (1.3 cm) in diameter. These polyps can reproduce by budding but remain attached to each other, forming a multipolyp colony with a common skeleton, which may be up to several meters in diameter or height, according to species. Reef growth is slow; an individual colony may grow from 0.5 to 7 inches (1 to 18 cm) per year, depending on the species. There are more than 45 species of hard corals found in Florida.

**Soft corals** (octocorals) have eight (or multiples of eight) tentacles on their bodies. Their outer crusts contain small spikes of calcium carbonate, while their inner cores may contain gorgonin, a flexible, fibrous wood-like protein. This gives the soft coral colony the ability to flex with the ocean waves and currents. Most soft corals form colonies that take on a variety of forms such as sea fans, sea whips, and others. Some soft corals may host symbiotic algae. In Florida, there are 35 species of soft coral.

**Sponges**, like corals, are immobile aquatic invertebrate animals, but they are otherwise completely different organisms with distinct anatomy, feeding methods, and reproductive processes. Nevertheless, marine sponges are also important organisms within the coral reef community, and more than 70 species can be found along the Florida Reef Tract.

## Nature Is Amazingly Complex!

This chapter has just scratched the surface of the tremendous complexity of coastal ecosystems. While humans have, over the millennia, greatly modified our planet's natural ecosystems, we are not separate from nature; instead, we are embedded into it and are totally dependent upon it. Chapter 2 covers some of the ways humans have modified Florida's coastal ecosystems over the years—both positively and negatively.

## Notes

1. United Nations Economic Commission for Europe (UNECE), "Ecosystem Services," accessed January 28, 2025. https://unece.org/ecosystem-services-0.
2. Florida Fish and Wildlife Conservation Commission (FWC), "Mangrove Forests," accessed January 28, 2025. https://myfwc.com/research/habitat/coastal-wetlands/mangroves/.
3. Bilkovic et al., *Review of Boat Wake Wave Impacts;* Shuster, et al., "Erosive Potential of Recreational Boat Wakes."
4. NOAA, *Patterns and Projections of High Tide Flooding.*
5. Myers and Ewel, *Ecosystems of Florida*, 436.
6. NOAA, "What Is a Mangrove Forest?," accessed January 28, 2025. https://oceanservice.noaa.gov/facts/mangroves.html.
7. Robert et al., "Mangrove Growth Rings."
8. Myers and Ewel, *Ecosystems of Florida.*
9. Wooller et al., "A Multiproxy Peat Record."
10. Myers and Ewel, *Ecosystems of Florida.*
11. Osland et al., "The Impacts of Mangrove Range Expansion."
12. Myers and Ewel, *Ecosystems of Florida.*
13. Bilkovic et al., *Living Shorelines,* 217.
14. Chesapeake Bay Foundation, "Oyster Fact Sheet," accessed January 28, 2025, https://www.cbf.org/about-the-bay/chesapeake-wildlife/eastern-oysters/oyster-fact-sheet.html.
15. Volety, "Habitat: Oyster Reefs."
16. Kitsikoudis et al., "*In-Situ* Measurements of Turbulent Flow."
17. Seavey et al. "Decadal Changes in Oyster Reefs."
18. van Tussenbroek, et al. "Experimental Evidence of Pollination in Marine Flowers by Invertebrate Fauna."
19. Borum et al., "Oxygen Movement in Seagrasses."
20. EPA, "Basic Information About Coral Reefs," accessed January 28, 2025, https://www.epa.gov/coral-reefs/basic-information-about-coral-reefs.

## References

Baker, Andrew, Craig J. Starger, Tim R. McClanahan, and Peter W. Glynn. "Corals' Adaptive Response to Climate Change." *Nature* 430, no. 741 (2004). https://doi.org/10.1038/430741a.

Bilkovic, Donna, Molly M. Mitchell, J. Davis, E. Andrews, et al. *Review of Boat Wake Wave Impacts on Shoreline Erosion and Potential Solutions for the Chesapeake Bay.* STAC, no. 17–002. Edgewater, MD, 2017. http://ccrm.vims.edu/2017_BoatWakeReviewReport.pdf.

Bilkovic, Donna Marie, Molly M. Mitchell, Megan K. La Peyre, and Jason David Toft, eds. *Living Shorelines: The Science and Management of Nature-Based Coastal Protection.* CRC Press, 2020.

Borum, J., K. Sand-Jensen, T. Binzer, O. Pedersen, and T. Greve. "Oxygen Movement in Seagrasses." In *Seagrasses: Biology, Ecology and Conservation,* edited by Anthony W. D. Larkum, Robert J. Orth, and Carlos M. Duarte, 255–70. Springer, 2006.

Florida Department of Environmental Protection. "Office of Resilience and Coastal Protection." Accessed January 28, 2025. https://floridadep.gov/rcp.

Florida Fish and Wildlife Conservation Commission. "Seagrass Research." Accessed January 28, 2025. https://myfwc.com/research/habitat/seagrasses/.

Florida Sea Grant. "Science Serving Florida's Coasts." Accessed April 4, 2025. https://www.flseagrant.org/.

Havens, Karl, Gary Lytton, and William Seaman. *A Practical Guide to Estuary-Friendly Living.* Florida Sea Grant. University of Florida, SGEB-64, 2012. https://www.flseagrant.org/publication/a-practical-guide-to-estuary-friendly-living/.

International Union for Conservation of Nature (IUCN). "Peatlands and Climate Change," Issues Brief, December 2021. https://iucn.org/resources/issues-brief/peatlands-and-climate-change.

Kitsikoudis, V., K. M. Kibler, and L. J. Walters. "In-Situ Measurements of Turbulent Flow over Intertidal Natural and Degraded Oyster Reefs in an Estuarine Lagoon." *Ecological Engineering* 143 (2020): 205688.

McIvor, Anna, Iris Möller, Tom Spencer, and Mark Spalding. *Reduction of Wind and Swell Waves by Mangroves.* Report No. 1: Natural Coast Protection Series. The Nature Conservancy and Wetlands International, 2012, https://www.conservationgateway.org/ConservationPractices/Marine/crr/library/Documents/wind-and-swell-wave-reduction-by-mangroves.pdf.

McIvor, Anna, Tom Spencer, Iris Möller, and Mark Spalding. *The Response of Mangrove Soil Surface Elevation to Sea Level Rise.* Report No. 3: Natural Coastal Protection Series. The Nature Conservancy and Wetlands International, 2013. https://www.conservationgateway.org/ConservationPractices/Marine/crr/library/Documents/mangrove-surface-elevation-and-sea-level-rise.pdf.

Myers, Ronald L., and John J. Ewel, eds., *Ecosystems of Florida.* University of Central Florida Press, 1990.

NOAA. Office for Coastal Management. Accessed January 28, 2025. https://coast.noaa.gov/states/florida.html.

NOAA. *Patterns and Projections of High Tide Flooding Along the U.S. Coastline Using a Common Impact Threshold.* National Oceanic and Atmospheric Administration, 2018. https://tidesandcurrents.noaa.gov/publications/techrpt86_PaP_of_HTFlooding.pdf.

Osland, Michael J., et al. "The Impacts of Mangrove Range Expansion on Wetland Ecosystem Services in the Southeastern United States: Current Understanding, Knowledge Gaps, and Emerging Research Needs." *Global Change Biology* 28, no. 10 (2022): 3163–87.

Robert, Elizabeth, Nele Schmitz, Judith Okello, and Ise Boeren. "Mangrove Growth Rings: Fact or Fiction?" *Trees* 25, no. 1 (2010):49–58. https://www.researchgate.net/publication/226008257_Mangrove_growth_rings_Fact_or_fiction.

Seavey, J. R., W. E. Pine III, P. Frederick, L. Sturmer, and M. Berrigan. "Decadal Changes in Oyster Reefs in the Big Bend of Florida's Gulf Coast." *Ecosphere* 2, no. 10 (2011): 1–14.

Shuster, Robert, Douglas J. Sherman, Mark S. Lorang, Jean T. Ells, and Frank Hopf. "Erosive Potential of Recreational Boat Wakes." Special issue, *International Forum for the Littoral Sciences*, no. 95 (2020): 1279–83, https://www.jstor.org/stable/48748894.

Smithsonian: National Museum of Natural History. "Ocean: Find Your Blue." Accessed January 28, 2025. https://ocean.si.edu.

South Carolina Department of Natural Resources. "Dynamics of the Salt Marsh." *Sea Science*, 2014. https://www.dnr.sc.gov/marine/pub/seascience/dynamic.html.

van Tussenbroek, B., N. Villamil, J. Márquez-Guzmán, R. Wong, L. Monroy-Velazquez, V. Solis-Weiss. "Experimental evidence of pollination in marine flowers by invertebrate fauna." *Nature Communications* 7, 12980 (2016). https://doi.org/10.1038/ncomms12980

Volety, Aswani K. "Habitat: Oyster Reefs." *MARES—Marine and Estuarine Goal Setting for South Florida* (September 2013): 67–79. https://www.aoml.noaa.gov/ocd/ocdweb/docs/MARES/MARES_SWFS_ICEM_20130913_Appendix_OysterReefs.pdf.

Wooller, M. J., R. Morgan, S. Fowell, H. Behling, and M. Fogel. "A Multiproxy Peat Record of Holocene Mangrove Palaeoecology from Twin Cays, Belize." *Holocene* 17, no. 8 (2007): 1129–39, https://doi.org/10.1177/0959683607082553.

# 2

# People and Florida's Coastal Zones

**THE BEGINNING OF THIS** chapter highlights some of the problems caused by too much development in fragile habitats and by unsustainable landscaping practices that cause harm to coastal ecosystems; but the second half of this chapter, and indeed the remainder of the book, provides strategies to mitigate the damage and to build more resilient shorelines. The scale of the issues may seem daunting, but with proactive and intentional action, it is possible to reverse environmental harm.

Florida is faced with many challenges, and among the most serious is the increased coastal development combined with accelerating sea level rise. Although the rate of sea level rise varies geographically, the National Oceanic and Atmospheric Administration (NOAA) tidal station data from Key West and Virginia Key indicate that sea levels have increased 8 inches (20.3 cm) since 1950; in addition, the speed of sea level rise has accelerated to 1 inch (2.5 cm) every three years. These problems are not limited to the southern regions of Florida but have been evident across the whole state.[1]

Florida has more than eighty unique ecosystems and 8,460 miles (13,615.1 km) of tidal shoreline. But large parts of the state have been subjected to rampant development without regard to its effects on the natural world. Salt marsh, mangrove, and oyster habitat losses are extensive. For example, according to the Florida Natural Areas Inventory (2010), between 1950 and 1980, nearly half the salt marsh in the Charlotte Harbor and Tampa Bay area was destroyed by development activity; and in the Indian River Lagoon, roughly 70 percent of mangrove areas have been lost due to mosquito control impoundments built for human use.[2] On

Florida's west coast, Naples Bay has lost 80 percent of its oyster reefs since 1950, while oyster reef habitat has decreased by 99 percent in southeast Florida.[3]

## Population Growth and Development

As of this writing, more than one thousand people move to Florida every day.[4] To look at this in a different way, Florida's population grew 18 percent from 2010 to 2022, starting with 18.8 million people in 2010. For comparison, the population in the entire United States grew only 7.7 percent during that same period.[5] So the sheer number of people, with all their cars, ATVs, boats, and their mostly single-family houses with large, highly maintained lawns and mostly nonnative landscaping, puts an enormous strain on Florida's resources in many ways. (See chapter 9 for more information on lawns and alternative lawn care strategies.)

**FIGURE 2.1** Aerial view of Cape Coral, Florida, showing an extensive canal system and the resulting waterfront properties. Photo by David Wilson, 2017, Cape Coral, Florida *(https://www.flickr.com/photos/davidwilson1949/44302281040/; reproduced without modifications under creative commons license CC BY 2.0 https://creativecommons.org/licenses/by/2.0/).*

Development has continued unabated for decades to accommodate this influx of people, and in many areas of Florida, waterfront lots have been created by filling in wetland areas as well as by digging canals to maximize the number of waterfront properties. In many of these created waterfront lots, bulkheads or other shoreline armoring have been installed to maximize the land. (See chapter 3 for more information on the problems of armoring.) In the city of Cape Coral alone, where the population growth was 26 percent from 2010 to 2020 according to census data, there are 400 miles (643.7 km) of canals;[6] and this pattern is repeated many times over in Florida, mostly in the southern half of the peninsula.

Water quality within residential canal systems is often very poor, due to limited flow through the artificial waterways, which leads to water stagnation. This issue is further exacerbated when runoff from adjacent landscapes carries organic material such as grass clippings and other nutrient-rich material (e.g., fertilizers and pet waste) into the water, further fueling the growth of often undesirable algal species. Decaying organic debris and dead algal material leads to very low oxygen levels in the water that can in turn cause problems such as fish kills. Over time, there can be a buildup of unsightly and smelly muck material on the bottom of the canals, necessitating costly dredging and waste disposal activities.

**FIGURE 2.2** Canal-front development in Punta Gorda, Florida. Canals like this are common in much of the southern half of coastal Florida. Note the lawns extending right to the edge of the bulkheads, which results in too many nutrients washing into the water from lawn fertilizer applications, grass clippings, and other organic matter. There is often limited water flow in canals, which leads to algae blooms caused by excess nutrients. *Photo by Florida Sea Grant.*

Regardless of whether a waterfront property is part of a canal system, the aggressive management of lawns near bodies of water is a major source of pollution, as it includes eroded soil, pesticides, grass clippings, leaves, and nutrients from fertilizers that flow from the landscapes during rain events, especially during heavy rains. (See chapter 9 for a more complete discussion of lawns and suggestions for more sustainable care.)

Also, as soon as it starts to rain, the stormwater washes over streets, parking lots, and other impervious (paved) surfaces and carries substances such as pet waste, chemicals, drippage from vehicles, spilled gas, and waste from washing cars and pressure washing paved driveways in addition to lawn chemicals and organic yard waste. Unfortunately, stormwater is usually not routed to a treatment facility but rather flows overland or into storm drains that convey it directly out to natural bodies of water, along with all of the pollutants it carries. This combination of stormwater from landscapes and streets flowing into bodies of water is referred to as nonpoint source pollution. (For more information, see sidebar "Point Source and Nonpoint Source Pollution.")

It is this nonpoint source pollution that we, as individuals and communities, have the most control over, and we have many real opportunities to reduce its

**FIGURE 2.3** Stormwater runoff from roads carries petrochemicals from vehicle drippage and other chemical spills. This liquid drains into bodies of water; it is an example of nonpoint source pollution. *Photo by Alex Filim/Shutterstock.com.*

effects on our waterways. (See chapters 8 and 9 for ideas on implementing shoreline friendly measures, including sequestering stormwater and installing plants that act as buffers to reduce nonpoint source pollution.)

## Plantings on Shoreline Properties

Another factor in the default waterfront landscapes installed by developers is that most of the plants are not native to Florida; in fact, many of them are known invasive species in Florida. (See sidebar "Florida's Native, Nonnative, and Invasive Plants.") Research by Doug Tallamy, a professor of entomology at the University of Delaware and the author of several books, has shown that even small landscapes planted with native plants can make a significant difference in restoring the local working ecosystem. Native plants provide food for insects, which provide food for birds; Tallamy's research shows that one pair of Chickadees (*Poecile* spp.) needs 6,000 to 9,000 caterpillars to raise one clutch of young. Chickadees lay five to eight eggs at a time. The hosts for those caterpillars (butterfly and moth larvae) are almost always native plants—mostly trees. While the Chickadee is not a Florida bird, Tallamy's work has been cited in studies of bird habitat in urban and suburban areas of Florida, where there are similar findings and outcomes.[7]

In addition to their importance in restoring good habitat, native plants, once they are established, thrive in Florida's climate with its five-month wet season and seven-month dry season without much additional irrigation. Native plants also generally do well in unamended soils. This means that homeowners and other landscape managers who grow mostly native plants will use less fertilizer, use less water, and require less ongoing maintenance in the long run. Landscapes that are filled with mostly native plants will support birds, pollinators, and other wildlife; and their minimal care requirements will also cause less pollution of the wetlands and waterways, all while saving money for the homeowner or property manager.

In this book, mostly native plants are recommended for shoreline properties, upland buffers, and living shorelines, because rebuilding native habitat is an important goal for establishing resilient landscapes in a changing world.

## POINT SOURCE AND NONPOINT SOURCE POLLUTION

The US Environmental Protection Agency (EPA) defines *point source pollution* as "any single identifiable source of pollution from which pollutants are discharged, such as a pipe, ditch, ship, or factory smokestack." Factories, including oil refineries; pulp and paper mills; and chemical, electronics, and automobile manufacturers typically expel one or more pollutants in their discharged waters (called effluents). Some factories discharge their effluents directly into nearby bodies of water, while others treat it on-site before it is released; still others send their waste to sewage treatment plants to be processed before being released. Sewage treatment plants also treat human waste before releasing treated effluent to a wetland or body of water.

Starting in 1948, the US government implemented the Federal Water Pollution Control Act to regulate point source pollution, but it was significantly reorganized and expanded in 1972 when it was renamed the "Clean Water Act." It has been updated since then and, as of this writing, the most recent update was in 2022. These regulations apply to the identifiable point sources of pollution.

**Nonpoint source pollution** begins when it rains and the rain hits impervious surfaces such as roofs and streets, or pervious surfaces such as lawns and flows across streets, parking lots, yards, construction sites, farm fields, landfills, and golf courses. Along the way, the rainwater picks up litter, oil, grease, metals, rubber, dirt, fertilizers, pesticides, animal waste, and other pollutants left behind by people, animals, automobiles, and other machinery. These pollutants mix with the rain and flow into waterways such as rivers, streams, lakes, wetlands, estuaries, and even into the groundwater. The term *nonpoint source pollution* is derived from the concept that there is no single point from which the pollution comes; it comes from everyone and everywhere. Nonpoint source pollution is the nation's and the state's number one threat to water quality, and unlike point source pollution, few regulations affect it.[a]

These pollutants harm the aquatic ecosystems, but even pollutants that seem like they would be less harmful such as nutrients and sediments can cause significant problems.

- **Nutrients**: An overabundance of nutrients usually causes the overfertilization of wetlands or bodies of water that leads to excess algae growth. During the day, when algae is photosynthesizing, the algae produce more oxygen in the water than they take in as they respire. At night, when no oxygen is being produced, the algae continue to respire, like all the other aquatic organisms. But when there is too much algae, this nighttime uptake of oxygen may cause organisms in that region of the water to die, including the algae itself. In addition, when the algae die, they are broken down by bacteria, and the bacteria also consume oxygen. The areas where this happens are called dead zones or temporary dead zones.

**FIGURE 2.4** A 2016 fish kill in the Banana River, Cocoa Beach, Florida, was the result of a nutrient-driven algae bloom. *Photo by Marjorie Shropshire.*

- **Sediment**: Eroded soil from yards, farm fields, construction sites, and shorelines is carried by water as sediment. Sediments can also carry other pollutants such as metals and toxic chemicals. When it reaches a body of water, it causes the water to become cloudy, which reduces photosynthesis, which in turn reduces oxygen content in the water. In addition, loss of aquatic plants like seagrass that hold sediment in place can make turbidity worse, since sediment gets stirred up in the water by the action of wind, tides, and boat wakes. Sediment makes it difficult for aquatic organisms to see and to feed properly. In addition, it can damage fish gills and the breathing organs of other aquatic animals. Fish spawning habitat covered by sediment reduces reproduction rates. These are just some of the problems caused by too much sedimentation.[b]

[a] Environmental Protection Agency, "Polluted Runoff: Nonpoint Source (NPS) Pollution," accessed January 31, 2025, https://www.epa.gov/nps.

[b] Aswani Volety and Vincent G. Encomio, *Biological Effects of Suspended Sediments on Shellfish in the Charlotte Harbor Watershed: Implications for Water Releases and Dredging Activities,* Final Report to the Charlotte Harbor National Estuary Program (Florida Gulf Coast University, 2006), https://sarasota.wateratlas.usf.edu/upload/documents/SedimentsShellfishCaloosa_FGCU.pdf.

## FLORIDA'S NATIVE, NONNATIVE, AND INVASIVE PLANTS

Over the millennia, plants, insects, birds, soil inhabitants, and other critters have adapted to depend upon each other in many ways. In a natural, native ecosystem, everything works together so that sources of food are available to correspond to life cycles of the wildlife in that ecosystem. For instance, while most birds feed on insects during most of the year, in the fall, many plants produce seeds and berries that feed the birds through winter when the supply of insects may be reduced.

Over the course of history, humans have drastically altered the landscapes by clearing vast areas for agriculture and development, which has led to habitat fragmentation (creating "islands" of native ecosystems surrounded by development). This lack of connectivity has contributed to the loss of biodiversity, making it more difficult for plants and animals to successfully complete their life cycles.

In addition, humans have imported nonnative plants for food, to stop erosion, for screening, for ornamental uses in gardens, and as houseplants. Also, some plants were imported by accident, together with other plants or plant-based products. While most plants from other parts of the world are not particularly harmful to our natural areas, some are aggressive without their natural predators (insects and other herbivores from their native regions) to keep them in check. Many have become too abundant, replacing significant populations of native plants and upsetting the natural balances in our native ecosystems.

The plants that have caused the most damage to native ecosystems, as determined after years of study and monitoring, have been labeled as **invasive** by the Florida Invasive Species Council (https://floridainvasivespecies.org). As of this writing, there are about 170 invasive plant species in Florida. The problem with invasives is not trivial, and the problems with many invasive plant species are increasing. The state of Florida and various environmental organizations spend many millions of dollars and thousands of man-hours each year—both paid and volunteer—removing invasive plants. One study estimated that the Florida government spends $45 million/year on removing invasive plants, and this does not include the spending and labor of private organizations; and despite these problems, some of these invasive plants are still being sold.[a]

Nonnative plants, which have not evolved with our local ecosystems, may not play appropriate roles in feeding and hosting our native insects or birds. And some, such as the invasive, but widely sold, heavenly bamboo (*Nandina domestica*) can poison birds, especially migrating birds, notably Cedar Waxwings (*Bombycilla cedrorum*), that gorge on the berries to gain energy for their migration flight.

Brazilian peppertree (*Schinus terebinthifolia*) is an example of a plant that was imported from South America to serve as a fast-growing shrub for windbreaks or screening. It is sometimes called the Florida holly since it has attractive red berries that are sometimes dried and used as a pepper substitute. It is in the poison ivy family (Anacardiaceae) and the plant contains some of the same oil (urushiol) as poison ivy that gives people rashes, so controlled burning cannot be used as part of its control.

**FIGURE 2.5** The South American Brazilian peppertree (*Schinus terebinthifolia*) was brought to Florida in the mid-1800s as an ornamental plant because of its bright red berries and evergreen foliage. Birds eat the red berries and deposit the seeds great distances from the original plant. It is now one of Florida's 170 invasive plants. *Photo by B. Navez, 2006, Wikimedia Commons (https://commons.wikimedia.org/wiki/File:Schinus_terebinthifolius_fruits.JPG; reproduced without modification under creative commons license CC BY-SA 3.0. https://creativecommons.org/licenses/by-sa/3.0/deed.en).*

It has become horribly invasive, especially in damp coastal areas where it often crowds out everything else; its numerous branches can be upright, reclining, or almost vine-like. Just cutting it off does not kill it either; it will grow back even more vigorously. Brazilian pepper can infiltrate mangrove forests. Floating seeds pushed by waves and boat wakes can move into the intertidal zone, where they grow and can outcompete the mangroves.[b]

All sustainably managed landscapes or wetlands should include regular monitoring for the removal of invasive plants. Even if an invasive plant seems tame in the landscape, if it is growing in a container, or even if it is a "sterile hybrid," it may still be cross pollinating with uncontrolled populations and adding to the vitality of the local population of invasive species. Seeds can always be carried by birds and other animals or float on water to new locations.

[a] Drew Hiatt, Kristina Serbesoff-King, Deah Lieurance, Doria R. Gordon, and S. Luke Flory, "Allocation of Invasive Plant Management Expenditures for Conservation: Lessons from Florida, USA," *Conservation Science and Practice* 1, no. 7 (2019): e51, https://conbio.onlinelibrary.wiley.com/doi/10.1111/csp2.51.

[b] Donna Marie Bilkovic, Molly M. Mitchell, Megan K. La Peyre, and Jason David Toft, eds, *Living Shorelines: The Science and Management of Nature-Based Coastal Protection* (CRC Press, 2020), 236.

## Estuaries

Although estuaries are vitally important to the economic and ecological health of Florida, they have been severely damaged due to over-development, poor planning, and people not realizing what impacts their activities have on estuarine ecosystems. Most of the damage is loss of habitat due to development, where the watersheds and shorelines of estuaries have been impacted by coastal communities, highways, harbors, and marina facilities. With this development comes pollution. Fertilizers, organic materials, eroded soils, oil, gas, sewage, pharmaceuticals, and other pollutants from residential, agricultural, and industrial activities all end up in our estuaries. Humans have also drastically modified the dynamics of freshwater delivery to many estuaries through activities such as the creation of drainage canals, the construction of reservoirs/dams, and the pumping of groundwater that would otherwise have served as stream baseflow (that portion of the streamflow that is sustained between precipitation events).

Changes in freshwater delivery to the coast result in mismatches in supply and need and timing, allowing too much or too little water to reach the coastal ecosystems, often also at the wrong time of year. This also causes damage to estuaries and may have serious impacts on the life cycles of fishes, crabs, shrimp, and other organisms. A good example of such damage is the Lone Cabbage Reef in the Suwannee River estuary, where the withdrawal of groundwater for urban, residential, and agricultural use has decreased the amount of freshwater reaching the coast. This decline in freshwater delivery has been linked to large-scale die-offs of oysters and the subsequent degradation of vast oyster reef systems in the Big Bend region of Florida.[8]

Another example of estuary damage is the Indian River Lagoon, where the construction of drainage canals has artificially increased the watershed size by approximately 2.5 times (from 572,000 acres to 1.4 million acres).[9] This means that stormwater—and the pollution contained within it—drains from a much larger area into the lagoon; and therefore the overall pollutant load has drastically increased. The issue is worsened by the fact that drainage canals allow water to exit the landscape very quickly, and therefore the processes that would normally filter and trap pollutants, such as percolation through soils or capture and retention in vegetated wetland landscapes, are greatly reduced. This allows a greater quantity of more polluted water to flow into the lagoon, where it can cause severe water quality issues.

## Beaches and Dunes

Along high-energy ocean shorelines, beaches and dune systems are common. In these dynamic ecosystems, sand continuously shifts with the tides and other currents. In a completely natural condition, beaches are maintained by sand being transported from up-current areas through a process called longshore drift, a sort of natural conveyor belt that continually moves sand along the coastline. Areas of erosion and sand deposition shift over time, but as long as the upstream source of sand, often a river, continues to supply sand, the beach width will generally remain stable. Occasionally, storms also shift larger quantities of sand. In natural dune and beach systems, this constant movement of sand is not a problem.

Beaches are dynamic, but buildings, bridges, and roads are static. Problems occur when these static structures are built on dynamic, moving beaches and dune ecosystems. These problems are exacerbated when erosion rates are fueled by human practices such as shoreline armoring, the damming or diversion of rivers, and dredging; these practices disrupt the natural conveyor belt of sand, essentially starving downstream beaches of the sand supply necessary to maintain them. If buildings and roads were not built so close to the shore and if coastal

**FIGURE 2.6** This seawall was constructed to protect property built too close to the shoreline. Seawalls increase erosion and eliminate the protective dune systems naturally found at the backs of the beaches. *Photo by Marjorie Shropshire.*

modifications that disrupt sand supplies were avoided, we would not have to worry about shoreline structures falling into the sea.

Many communities try to maintain their beaches by adding sand each year (a practice termed *nourishment*), but this is a huge expense and rarely works for long because the currents and other conditions that washed the beach away still exist. Plus, this imported sand can be detrimental to local beach life if its color, grain size, and other properties are too different from the local sand type. There can also be environmental impacts related to the processes used to mine and transport the sand (i.e., offshore dredging or large-scale trucking). Jetties and groins are structures used in an attempt to keep sand in place, which they do to some degree, but their use comes at the expense of other beach areas farther downstream from the prevailing longshore drift.

By building structures too close to the beach, people have been creating a coastal squeeze. That is, the dynamic shorelines do not have enough space to shift without risking coastal buildings, roads, and other infrastructure. As erosion in front of armored structures progresses, elevations sink, and this often squeezes out portions of upper beach/foredune systems that are critical areas for sea turtle nesting and more.

## Mangrove Forests

The Florida Department of Environmental Protection (FDEP) estimates that mangrove forests contribute $7.6 billion annually to the state's economy and create 109,000 jobs in Florida. Too many mangrove forests have been removed from the edges of Florida's waterways over decades of development, often just to gain a water view. As a result, our shorelines are more vulnerable to tropical storm damage, and our native bird and fish populations are at higher risk of steep decline, especially after stressful events such as hurricanes or red tides. In addition, the removal of mangroves results in the release of a huge amount of carbon into the atmosphere that had been stored underground—sometimes for thousands of years—in the form of peat.

To protect the environment and prevent these harmful effects, the Florida Mangrove Trimming and Preservation Act regulates trimming and alteration of mangroves. (See chapter 5.) Once most mangrove trees are cut down, they are unlikely to grow back from the roots; but of the different mangrove species, black mangroves are more likely to grow back after damage to the tree has occurred from cutting, frost, or other events.

## Salt Marshes

Salt marsh systems are naturally dynamic and constantly changing. However, upland and coastal development has greatly impacted Florida's salt marshes, especially historically, before the enactment of protective regulations. The adverse impacts of urban development of tidal marshes include the degradation of water quality, the filling of marshes, their impoundment for mosquito control, and increased rates of erosion as well as the impacts of other alterations such as bulkheading and alterations in natural sediment supply.

The elimination and alteration of Florida salt marshes has negatively affected fishery resources, harmed water quality, and left many coastal uplands exposed to direct impacts from storms.

The benefits of saltmarsh ecosystems make them a practical choice for inclusion in mitigation and adaptation approaches, because marshes occupy much of the same low lying coastal areas that are especially vulnerable to sea level rise.

## Seagrass Beds

As is the case for mangroves and salt marshes, many acres of seagrass beds also have been destroyed by human activities.[10] Although seagrass is not commonly cultivated in Florida, its economic value can be measured through other industries, such as commercial and recreational fisheries and nature and wildlife tourism that rely on this habitat to survive. Since about 85 percent of Florida's fish species spend at least part of their life cycle within seagrass communities, seagrasses are vital to the survival of the state's fishing industries.[11]

There are multiple stressors that have contributed to the past and current declines of seagrass meadows in Florida. Chief among these are the excess loading of nitrogen and phosphorus, derived largely from nonpoint source pollution, and the highly modified watersheds. Excess nutrients fuel algal blooms that shade seagrasses. Algal blooms have driven seagrass declines in Florida Bay, Tampa Bay, and the Indian River Lagoon, among other coastal areas.[12] Elevated sea surface temperatures and boat-generated scarring are also major threats.[13] Many seagrass beds in Florida have also been lost through the creation of canalized waterfront developments, as described above. In many cases, seagrass-covered areas were dredged to the point that even restoration efforts will not be able to reestablish seagrasses in these areas, as the water in the dredged areas is too deep for seagrasses to survive and canal water quality is often too poor even in shallow canals. Once lost, seagrass meadows are often difficult to reestablish because fragmented or sparse seagrass is more vulnerable to being wiped out from smaller-scale stressors; fully established meadows can be quite resilient to stress. Therefore, conservation of our remaining seagrass meadows is critical.

## Coral Reefs

Pollutants, nutrients, sediments, and litter enter near-shore waters via rivers, streams, underground seepage, wastewater, and stormwater runoff. Coral ecosystems, even those that are hundreds of miles from the coast, can be affected by such pollutants, which reduce the clarity and quality of the water. It does not matter how far-removed pollutants may seem, they all flow downstream, where they impact marine ecosystems, including coral reefs. (See sidebar "Point Source and Nonpoint Source Pollution.")

## Ecosystem Damage

Unfortunately, whether through short-sightedness, lack of understanding, or despite the consequences, human activities have, on balance, resulted in widespread degradation and loss of natural ecosystems. Because of their natural resilience, the services the ecosystems provided to humans seemed endless. Over time, though, all this human-created damage is becoming more and more evident; we, as a civilization, have injured ourselves by disregarding the services the ecosystems provide for us. (See sidebar "Ecosystem Services.")

## People *Do* Care About Florida's Coastal Areas

By contrast, there are people who have also preserved, conserved, and restored native habitats. Organizations such as the Florida Native Plant Society, the Florida Wildflower Foundation, the Florida Association of Native Nurseries, the Florida Wildflowers Growers Cooperative, the UF/IFAS Florida-Friendly Landscaping program, and other groups are working to educate homeowners and property managers about Florida's native plants. These groups also organize native plant sales to help more Floridians choose native landscaping plants.

## ECOSYSTEM SERVICES

An ecosystem is a functioning and interrelated combination of plants, animals, fungi, bacteria, and other organisms that exist together in a substrate where they depend upon each other to survive. The term *ecosystem services* describes how ecosystems benefit humans. The benefits can be direct or indirect—small or large. The Millennium Ecosystem Assessment, a major UN-sponsored effort to analyze the impact of human actions on ecosystems and human well-being, identified four major categories of ecosystem services: provisioning, regulating, cultural, and supporting services:[a]

1. **Provisioning Services** include fresh water, food (from plants and animals), medicines, building materials, fabrics, fuel, and more.
2. **Regulating Services** include processes that moderate natural phenomena. Plants clean air, capture carbon, cool the environment, filter water, and prevent erosion. Fungi, bacteria, and other organisms decompose wastes, which builds soil. Insects feed on plants, pollinate flowers, and are food for birds, which also disseminate seeds. All these processes work together to make ecosystems clean, sustainable, functional, and resilient.
3. **Cultural Services** can be traced back to the beginning of humankind, with ancient civilizations drawing pictures of animals, plants, and weather patterns on cave walls. A cultural service is a nonmaterial benefit that contributes to the development and cultural advancement of people born from interactions with nature including music, art, architecture, and recreation.
4. **Supporting Services** include natural processes, such as photosynthesis that produces oxygen gas and sugars, nutrient cycling, the creation of soils, and the water cycle. These processes allow the Earth to sustain basic life forms, including people.

Ecosystem services provided by coastal areas are particularly important to people, which is why we need to protect and restore them. For example, many of the fish we rely on for food spend at least part of their life cycles in coastal habitats. Wetlands absorb storm surges and control flood waters. Wetland plants absorb nutrients and chemicals from the water, and from runoff from uplands. They act as a natural filtration system Wetland plants and their soils store large amounts of carbon that, if released, would contribute to climate change. Wetlands are also a vital habitat for migratory birds, fish, and mammals, and their loss negatively impacts all these wildlife species.

To learn more, see the National Wildlife Federation's website and "Ecosystem Services" page at https://www.nwf.org/Educational-Resources/Wildlife-Guide/Understanding-Conservation/Ecosystem-Services.

[a] Neville Ash et al., eds, "Ecosystems and Their Services," in *Ecosystems and Human Well-Being: A Manual for Assessment Practitioners* (Island Press, 2010), 49–70, https://www.millenniumassessment.org/documents/document.300.aspx.pdf.

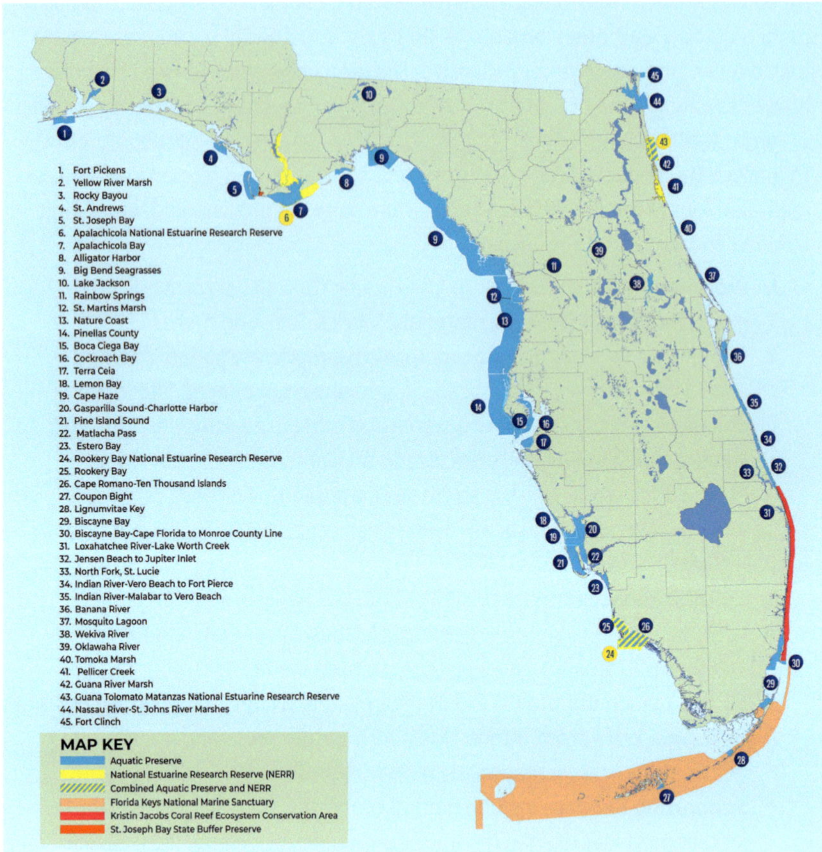


**FIGURE 2.7** A map of Florida's Aquatic Preserves, 2021. *Florida Department of Environmental Protection, Office of Resilience and Coastal Protection, Aquatic Preserve Program.*

## Florida's Parks and Conservation Areas

Florida has set aside many wonderfully diverse natural areas as parks and conservation areas. Many of them are in coastal areas, including 3 national parks, 3 national forests, 20 national wildlife refuges, 2 national seashores, 114 state parks, 5 state forests, 3 state reserves, and 19 state preserves. The Florida Park Service is the four-time winner of the Gold Medal honoring the nation's best state park systems, and it is one of the largest in the country, with nearly 800,000 acres (3,237.5 km$^2$) and 100 miles (160.9 km) of beaches set aside for public access. In addition, many of Florida's 67 counties have set aside and manage county parks, wetlands, and other natural areas. There are opportunities for Floridians to support many of these parks by volunteering, donating money, or by just visiting them, because the more visitors the parks host, the better the case they can make for more resources from the state and from local municipalities, counties, and environmental groups.

In 1975, Florida enacted the Aquatic Preserve Act. Aquatic preserves protect Florida's living waters to ensure they will provide habitat for bird rookeries and fish nurseries by protecting freshwater springs and salt marshes as well as seagrass meadows and mangrove forests. As of 2022, there were 42 preserves covering 2.6 million acres (10,521.8 $km^2$) mostly in coastal zones, but there are also some protected rivers and springs farther inland.

In 1998, The Florida Forever program was started where land is purchased to preserve natural and cultural heritage;[14] and 75 percent of Florida voters approved the Water and Land Legacy Act (WLLA) in 2014 to fund continued land acquisition.[15] So, most Floridians *do* care about preserving the best of Florida for wildlife and for future generations.

In 2021, the Florida Wildlife Corridor Act was signed into law with unanimous bipartisan support from the Florida legislature. The Florida Wildlife Corridor includes nearly 18 million acres of contiguous wilderness and working lands to protect wildlife and its movement around the state.[16]

## Florida's Environmental Organizations

Many Floridians join and support various environmental organizations that work to protect and preserve the "Real Florida" in various ways, including by working on the projects mentioned above and more. Below are the names and brief descriptions of some of the statewide organizations that are more nature oriented.

**Audubon Florida** has chapters in most of Florida's counties. The chapters hold meetings and organize field trips where people can learn about birds and Florida's bird habitats. The state organization also holds annual statewide meetings. (See https://fl.audubon.org/.)

**The Florida Native Plant Society** has more than thirty chapters. The chapters hold meetings, field trips, and plant sales and generally work on outreach about why native plants are so important. The state organization funds research on native plants, holds annual landscape awards, which recognizes native landscapes for both residential and commercial properties, holds state-wide conferences, and more. (See https://fnps.org/.)

**The Florida Wildflower Foundation** protects, connects, and expands native wildflower habitats through education, research, planting, and conservation. They run educational programs, provide grants for public wildflower projects, and more. (See https://flawildflowers.org/.)

**The Institute for Regional Conservation (IRC)** is dedicated to the protection, restoration, and long-term management of biodiversity on a regional basis as well as to the prevention of local extinctions of rare plants, animals, and ecosystems. It seeks to unite people and nature to restore our world. It focuses mostly on South Florida but intends to expand statewide. (See https://regionalconservation.org/.)

**The Florida-Friendly Landscaping Program** is an effort supported by the University of Florida's Institute of Food and Agricultural Sciences (UF/IFAS), a federal-state-county partnership. There are nine principles of Florida-Friendly Landscaping, and the ninth is to "protect the waterfront of both private and public properties." Here is their recommendation: "One of the most important steps you can take to protect any water body is maintaining a minimum 10-foot "low-maintenance zone" waterside. Do not mow, fertilize, or use pesticides in this zone. Protect native aquatic plants . . . and remove invasive exotic species." (See https://ffl.ifas.ufl.edu/.)

**The Surfrider Foundation** is an international organization whose mission is "dedicated to the protection and enjoyment of the world's ocean, waves, and beaches, for all people, through a powerful activist network." They have programs on plastic reduction and clean water, and they advocate for issues related to ocean conservation, beach access, climate change, and pollution. A Surfrider program that aligns well with living shorelines and Florida-Friendly Landscaping is their Ocean Friendly Gardens program, which encourages planting native plants, water conservation, and eliminating pesticides, herbicides, and commercial fertilizers in gardens to improve coastal water quality. There are multiple chapters of the foundation within Florida. (See: https://www.surfrider.org/).

**Waterkeepers Florida** is a regional organization of fifteen Waterkeepers in Florida. Each Waterkeeper oversees a river and watershed covering 45,000 square miles. Each Waterkeeper speaks for his or her region, working to protect and restore water quality through education, advocacy, and community engagement. You can find your local Waterkeeper by visiting https://www.waterkeepersflorida.org/.

**The Sierra Club** is one of the nation's largest and historic grassroots environmental organizations that has been working to protect wild places for more than 130 years. The Sierra Club Florida chapter works across the state to protect and promote the enjoyment of natural places in Florida. (See the national organization at https://www.sierraclub.org/; and the Florida chapter at https://www.sierraclub.org/florida).

**The Nature Conservancy** works throughout the world on conservation policy and science. Like the Sierra Club, their work is far-reaching and diverse, covering interests ranging from land and water conservation to climate change. But they have been particularly active in estuarine conservation and restoration (restoring oyster reefs and conserving living shorelines), throughout the United States, and especially in Florida. They support and work directly on projects that preserve and construct natural infrastructure to enhance the resiliency of Florida's coastlines. (See the national organization at https://www.nature.org/en-us/; and the state chapter at https://www.nature.org/en-us/about-us/where-we-work/united-states/florida/).

In addition to the local chapters of these national and statewide organizations, there are a number of independent environmental nonprofits. These groups work locally in their coastal areas to promote protection and restoration. Some of the groups that have been active in estuarine habitat and living shoreline restoration are as follows:

Brevard Zoo and its Restore Our Shores program, https://brevardzoo.org/ and https://restoreourshores.org/
Choctawhatchee Basin Alliance, https://basinalliance.org/
Ecosphere Restoration Institute, https://www.ecosphererestorationinstitute.org/
Florida Oceanographic Society, https://www.floridaocean.org
Marine Discovery Center, https://marinediscoverycenter.org/
Ocean Research and Conservation Association (ORCA), https://www.teamorca.org/
Oyster River Ecology Inc., https://oysterriverecology.org/
Sanibel-Captiva Conservation Foundation, https://sccf.org/
Sarasota Bay Watch, https://sarasotabaywatch.org/
St. Andrew Bay Watch, https://standrewbaywatch.org/
Tampa Bay Watch, https://tampabaywatch.org/

These groups have worked on living shorelines projects and are considered local experts. To find them and other professionals involved in living shorelines, visit their websites and the website of Florida Living Shorelines (https://floridalivingshorelines.com/), where more groups will be added to this list in the future.

## Government and Academic Institutions

There are several National Estuary Programs (NEPs) and local Estuary Programs throughout Florida that work to protect and restore water quality and ecological health in estuaries. The NEP is part of the Environmental Protection Agency (EPA) and was established under the Federal Clean Water Act, recognizing estuaries that are nationally significant. Currently, there are 29 NEPs across the United

States. Some local estuary programs may be in the process of receiving NEP status. Regardless, each program is place-based and works with local communities to ensure that their waters are swimmable and fishable for all, supporting science and restoration in each of their locations. Below is a list of the estuary programs in Florida.

Coastal and Heartland National Estuary Partnership, https://www.chnep.org/
Indian River Lagoon National Estuary Program, https://onelagoon.org/irlnep/
Pensacola and Perdido Bays Estuary Program, https://www.ppbep.org/
Sarasota Bay Estuary Program, https://sarasotabay.org/
St. Andrew and St. Joseph Bays Estuary Program, https://pc.fsu.edu/sasjbep
Tampa Bay Estuary Program, https://tbep.org/

The National Estuarine Research Reserves (NERRs, https://coast.noaa.gov/nerrs/) are coastal sites designated by NOAA for protection and research dedicated to understanding these coastal sites. They function as federal and state partners. NERR sites also work closely with their surrounding communities to achieve their aims. There are three NERR locations within Florida, with the FDEP serving as the state partner. Each NERR has its own living shoreline projects. To learn more about what each does, visit their web pages.

Apalachicola National Estuarine Research Reserve, https://coast.noaa.gov/nerrs/reserves/apalachicola-bay.html
Guana-Tolomato-Matanzas National Estuarine Research Reserve, https://coast.noaa.gov/nerrs/reserves/gtm.html
Rookery Bay National Estuarine Research Reserve, https://coast.noaa.gov/nerrs/reserves/rookery-bay.html

**Regional Estuarine Restoration Teams** (**ERTs**), created and convened by the Florida Fish and Wildlife Conservation Commission, are informal groups of restoration practitioners who plan and implement habitat restoration throughout Florida. These regional working groups represent partnerships between federal, state, private businesses, nonprofit organizations, and academic institutions working on estuarine restoration projects. They exist as forums to exchange ideas and information and to collaborate. Several of the regional ERT members are represented on Florida Living Shorelines (https://floridalivingshorelines.com/).

Several Florida colleges and universities are also engaged in restoration efforts within their regions. Their projects aim to restore estuarine habitats while educating

## CASE STUDY

### PROJECT #1. LAKE WORTH LAGOON / SNOOK ISLANDS PROJECT

**Location:** Palm Beach County, Lake Worth Beach, Lake Worth Lagoon

**Property Ownership:** Public

**Site History:** Historical dredging and filling going back to the 1800s and into the 1970s to construct the Lake Worth Golf Course destroyed shallow-water habitats and left a series of deep dredge holes along the 1.2 mile (1.9 km) shoreline. These holes had little to no ecological value, with poor water quality due to anoxic conditions as a result of silt and muck accumulating within the holes. By the late 1990s, the golf course shoreline was bulkheaded and began to fail, leading to erosion from wind-driven waves and boat wakes. Invasive species such as Australian pine, Brazilian pepper, and seaside mahoe colonized these areas, intermixing with native mangrove species (red, black, and white) fringing the shoreline.

**Construction Dates:** In 1998, Palm Beach County, in partnership with the US Army Corps of Engineers through the Section 1135 program, began project planning. The first phase of the Snook Islands Project began in 2003 and was completed in 2005. Phase I included the initial creation of the island and island habitat, with red mangrove (*Rhizophora mangle*) and smooth cordgrass (*Spartina alterniflora*) plantings and invasive species removal. In 2012, public use facilities were constructed, which included a boardwalk, kayak launch, fishing pier, and day-use docks to allow public access to the project area. Snook Islands Phase II began in 2013 with a partnership between Palm Beach County and the Florida Department of Transportation. Additional oyster reefs, mangrove areas, and seagrass habitat were constructed during Phase II. Between 2015 and 2018, Snook Islands Phase III saw the creation of the sixth and final island, an island with mudflats and two mounds for shorebird bird nesting, using dredged material acquired from two marina expansion projects.

**Living Shoreline Type:** Natural elements installed on constructed islands and shoreline, riprap to stabilize islands and serve as substrate for oyster settlement.

**Living Shoreline Building Blocks:** Sand; limestone stone rock to construct breakwaters; oyster reefs; red mangroves and cordgrasses were planted.

**Size:** The linear length of the project is 1.2 miles (1.9 km) along the golf course. The total size of the project is 118 acres (0.48 $km^2$ ), which includes 11 acres (0.04 $km^2$) of mangroves, 2.8 acres (0.01 $km^2$)of cordgrass marsh, and 2 acres (0.01 $km^2$) of oyster reefs. An additional 50 acres (0.2 $km^2$) of sea bottom was improved to facilitate seagrass recruitment.

**Permitting:** Individual South Florida Water Management District (SFWMD) Environmental Resource Permit and US Army Corps of Engineers (USACE) permit.

**Construction:** The dredge holes were filled with sediment to reconstruct bottom topography and establish optimal depths for intertidal and shallow subtidal vegetation. Sediment was also used to construct several islands.

**Phase I:** More than one million cubic yards of sediment were transported from a spoil storage site at Peanut Island ten miles (16 km) away. The islands were planted with red mangroves and cordgrasses. To stabilize the plantings and introduce substrate for oysters, 28,000 tons (25,401 tonnes) of limestone boulders were installed along the Snook Islands' perimeter, and mangrove planters were constructed along the shoreline. About 5 acres (0.02 $km^2$) of invasive exotic vegetation and approximately 1,800 linear feet (548.64 m) of the original concrete rubble from the collapsed seawall were removed from along the shoreline.

Red mangrove propagules were collected by volunteers and raised in a nursery before planting. Mangroves were planted via a series of volunteer-based planting events. Cordgrasses were planted by a contractor.

**Phase II:** In partnership with Florida Department of Transportation (FDOT) for advanced mitigation for bridges within Palm Beach County, 6 acres (0.02 $km^2$) of shallow-water habitat for seagrass recruitment, 0.39 acres (1,578 $m^2$)of mangrove and cordgrass habitat, and 0.14 acres (567 $m^2$)of oyster habitat were constructed.

**Phase III:** Dredged material was used to create the sixth and final island, approximately 1.7 acres (6,880 $m^2$) in size.

**Maintenance:** Yearly maintenance involves the removal of invasive plant species and the removal of white mangroves along the shoreline to maintain open vistas to the water and islands. The coastal bird mounds require bi-annual treatment and volunteer events pulling weeds to maintain sand areas for nesting birds. Monitoring to document initial and continuing performance maintenance is an important task as well.

**Monitoring:** Monitoring post-construction revealed that 80 percent of mangroves survived two years after planting. Cordgrass plantings coalesced in two years to create nearly 3 acres (12,141 $m^2$) of marsh but have since transitioned to mangroves (red, white, and black). Oysters began colonizing the rubble along the islands and mangrove planters. Surveys in 2006 and 2007 showed that seagrasses began colonizing the filled areas and increased to 14 acres (56,656 $m^2$). By 2010, seagrass acreage increased to 44.5 acres (178,062 $m^2$). Between 2010 and 2023 seagrass acres varied between no seagrass to 4.74 acres (19,020 $m^2$) in 2017. Florida Fish and Wildlife Commission began monitoring oysters in 2015 and fisheries in 2016.

**Notes:** The project is a good example of a large-scale living shoreline that is publicly accessible, like Project Greenshores in the Panhandle, and that serves as an example of how green infrastructure can be combined to create living shorelines to enhance resiliency and ecological health. Snook Islands Natural Area is within a highly urbanized waterway, the Lake Worth Lagoon, giving it a high recreational value for fishing and paddling. Areas of the project are accessible from land via boardwalks and a fishing pier. A kayak launch dock was part of this project. Information kiosks educate the public about living shorelines.

**Challenges:** The Lake Worth Lagoon is extensively developed, and the waterways experience a high degree of use from vessel traffic, emphasizing the need for innovative technologies like living shorelines to dampen wave energy and reduce shoreline erosion. Additionally, since this was a public project, maintenance responsibilities fall to the local government. The average annual vegetation maintenance cost is about $10,000 per year. Another challenge is finding areas in public ownership to enhance, otherwise approval of the upland riparian owner must be sought out. Additionally, dredge hole filling requires a lot of fill material, and finding a source and a location for staging a contractor's equipment is a challenge in popular urban locations. Shallow-water filling is more economical, but these areas tend to have resources to work around (existing seagrass, mangroves, cordgrass, oysters, or productive benthic communities) and are harder to permit.

**FIGURE 2.8** *(A)* The Snook Islands project looking north along the Lake Worth Lagoon. The living shoreline extends 1.2 miles (1.9 km) along the public golf course, and the project includes 11 acres of mangroves, 2.8 acres of cordgrass marsh, 2 acres of oyster reefs, and 50 acres of bottom that was improved to facilitate seagrass recruitment. *Snook Islands Natural Area, December 01, 2016. Photo by Palm Beach County Department of Environmental Resources Management. (B)* Breakwaters were constructed to shelter mangroves and cordgrass marsh from wave energy created by wind and boat wakes. The island was constructed with a sand bar and a deeper area to simulate natural marsh conditions. The deeper water between the sand bar and the breakwater was planted with mangroves, while the sandbar had both mangroves and cordgrass planted. *Photo by Ginny Stibolt.*

Thanks to Eric Anderson, Jena McNeal, Sabrina Fischer, and Jennifer Baez from PBC ERM for their input on this case study.

## REFERENCES

Engineering with Nature. "Snook Islands," May 6, 2021. https://ewn.erdc.dren.mil/built-projects/snook-islands/.

Palm Beach County, Environmental Resources Management. "Lake Worth Lagoon." Accessed January 31, 2025. https://discover.pbcgov.org/erm/Pages/Lake-Worth-Lagoon.aspx.

Palm Beach County, Environmental Resources Management. "Snook Islands." Accessed January 31, 2025. https://discover.pbcgov.org/erm/NaturalAreas/Snook-Islands.aspx.

Whitfield, Paula E., et al. *Engineering with Nature Principles in Action: Islands.* US Army Corps of Engineers, 2022. https://erdc-library.erdc.dren.mil/items/506e347d-bcc9-46d5-87d4-4b3a1e27ae0e.

students in environmental and ecological research. Many of these projects are also community-based, working with volunteers to install habitats. There are too many to list, and at the risk of omitting institutions deserving of recognition, it is a good idea to explore local individual college and university websites to learn more about their work. Several of the university faculty working on estuarine habitat and living shoreline restoration are also found on the Florida Living Shorelines website (https://floridalivingshorelines.com/).

## All Individual and Community Actions to Protect Florida's Coastal Zones Help

The following chapters in this book provide much more on specific actions that will have a huge collective impact on making our shorelines and our coastal ecosystems more resilient going forward. While each small project and improvement to one individual property may seem insignificant, if there are many thousands of individual projects, together they will make significant improvements to Florida's wonderful coastal ecosystems.

## Notes

1. See SeaLevelRise.org, the website of a group of scientists and managers studying sea level rise.
2. Dybas, "Florida's Indian River Lagoon."
3. Wallace et al., "Oyster Habitat Restoration."
4. Tampa Bay Economic Development Council, "How Many People Moved to Florida This Year?," March 29, 2023, https://tampabayedc.com/news/how-many-people-moved-to-florida-this-past-year/.
5. USA Facts. "Our Changing Population: Florida," last updated July 2022, https://usafacts.org/data/topics/people-society/population-and-demographics/our-changing-population/state/florida/.
6. Cape Coral, Florida, Public Works, "Contract Dredging Management," accessed February 1, 2025, https://www.capecoral.gov/department/public_works/canal_dredging.php.
7. See, for example, Google Scholar's list of publications by Mark Hostetler, a professor of Wildlife Ecology and Conservation at the University of Florida: https://scholar.google.com/citations?user=4it0F34AAAAJ&hl=en&oi=ao.
8. University of Florida, Wildlife Ecology and Conservation, "Recovery and Resilience of Oyster Reefs in the Big Bend of Florida," accessed February 1, 2025, https://wec.ifas.ufl.edu/oysterproject/problem.php.

9. Adams et al., *The Indian River Lagoon.*
10. Orth et al., "A Global Crisis for Seagrass Ecosystems."
11. Seaman, W. "Florida aquatic habitat and fishery resources."
12. Morris et al., "Seagrass in a Changing Estuary."
13. On elevated sea surface temperatures, see Carlson et al., "Sea Surface Temperatures and Seagrass Mortality." On boat-generated scarring, see Sargent et al. Scarring of Florida's Seagrasses.
14. Florida Department of Environmental Protection, "Florida Forever," accessed February 1, 2025, https://floridadep.gov/floridaforever.
15. Florida Department of Environmental Protection, Division of State Lands, "2022 Florida Forever Plan: Summary of Recommendations and Status as of December 2021," May 26, 2022, https://floridadep.gov/sites/default/files/2022_Executive_Summary_2.pdf.
16. See the Florida Wildlife Corridor Foundation at https://floridawildlifecorridor.org/.

## References

Adams, Amy W., Johnnie D. Ainsley, Derek S. Busby, Robert A. Day, Katherine A. Recore, and Troy B. Rice. *The Indian River Lagoon Comprehensive Conservation and Management Plan.* Indian River Lagoon National Estuary Program, Melbourne, FL, 1996. https://onelagoon.org/wp-content/uploads/irl_1996_ccmp.pdf.

Carlson, Daniel F., Laura A. Yarbro, Sheila Scolaro, Michael Poniatowski, Vicki McGee-Absten, and Paul R. Carlson Jr. "Sea Surface Temperatures and Seagrass Mortality in Florida Bay: Spatial and Temporal Patterns Discerned from MODIS and AVHRR Data." *Remote Sensing of Environment* 208 (2018): 171–88. https://www.sciencedirect.com/science/article/abs/pii/S0034425718300208.

Dybas, Cheryl Lyn. "Florida's Indian River Lagoon: An Estuary in Transition." *BioScience* 52, no. 7 (2002): 554–59.

Environmental Protection Agency. "Summary of the Clean Water Act: 33 U.S.C. §1251 et seq. (1972)." Accessed February 1, 2025. https://www.epa.gov/laws-regulations/summary-clean-water-act.

Florida Department of Environmental Protection. "Office of Resilience and Coastal Protection." Accessed February 1, 2025. https://floridadep.gov/rcp.

Galser, Evelyn, Angelikie Zafiris, Pablo L. Ruiz, and Franco A. C. Tobias. "Tracking Rates of Ecotone Migration Due to Salt-Water Encroachment Using Fossil Mollusks in Coastal South Florida." *Hydrobiologia* 569, no. 1 (2006): 237–57. https://www.researchgate.net/publication/225789125_Tracking_rates_of_ecotone_migration_due_to_salt-water_encroachment_using_fossil_mollusks_in_coastal_South_Florida.

Heimlich, Barry N., Frederick Bloetscher, Daniel E. Meeroff, and James Murley. *Southeast Florida's Resilient Water Resources: Adaptation to Sea Level Rise and Other Impacts of Climate*

*Change.* Florida Atlantic University, November 15, 2009. http://www.rcinef.org/SE_Florida_Resilient_Water_Resources.pdf.

Morris, Lori, Lauren M. Hall, Charles Jacoby, et al. "Seagrass in a Changing Estuary, the Indian River Lagoon, Florida, United States." *Frontiers in Marine Science* 8 (2022). https://doi.org/10.3389/fmars.2021.789818.

Orth, Robert J., et al. "A Global Crisis for Seagrass Ecosystems." *BioScience* 56, no. 12 (2006): 987–96.

Sargent, Frank J., Timothy J. Leary, David W. Crewz, and Curtis R. Kruer. *Scarring of Florida's Seagrasses: Assessment and Management Options.* FMRI Technical Report TR-1. Florida Department of Environmental Protection, Florida Marine Research Institute, St. Petersburg, FL, 1995.

Seaman, W. "Florida aquatic habitat and fishery resources." Florida Chapter of the American Fisheries Society. Eustis, Florida, USA. 1985.

Wallace, Charles, Edward Camp, and Ashley Smyth. "Oyster Habitat Restoration and Shoreline Protection." EDIS Publication FOR376, FR466. UF/IFAS, March 2022. https://edis.ifas.ufl.edu/publication/FR446.

# 3

# Introduction to Living Shorelines

## The History and Consequences of Shoreline Armoring

Shoreline armoring includes seawalls, bulkheads, riprap revetments (large rocks placed on the bank), and other structures constructed to protect property along the waterfront. Armoring is a stop-gap measure that can lead to accelerated erosion, habitat degradation, and the inability of a shoreline to adapt to changing stressors in the natural environment.[1] Humans have a long history of armoring shorelines against the erosive power of the ocean. Vertical seawalls are a common approach; they are designed to protect human-built infrastructure in coastal areas from erosion and storm surges.

The earliest known seawall, found in present-day Israel, dates back to approximately 7,000 years ago.[2] However, in Florida, most seawalls were built during the mid- to late twentieth century, particularly between the 1950s and 1970s. During this time, Florida experienced a major surge in coastal development and urbanization. The construction of seawalls was also driven by the desire to create more land for development, especially more waterfront property. A common practice was to dredge coastal areas and then backfill behind seawalls, creating instant waterfront property along man-made canals as described in chapter 2. This highly damaging process of "reclaiming" coastal areas is no longer allowed, but many of the seawall-lined canal communities built using this technique still exist today. Seawalls are common along many of Florida's riverine and beach shorelines, and new seawalls are being built all the time on existing waterfront or shoreline properties at great financial cost. But what other costs do seawalls have?

**FIGURE 3.1** Examples of armored shorelines. *(A)* A paved surface that extends to the edge of the high tide line at Cedar Key, Florida. *Photo by Savanna Barry. (B)* Concrete riprap dumped along the shore of Tampa Bay. *UF/IFAS file photo by Tyler Jones. (C)* Seawalls constructed to protect beachfront homes at Matanzas Inlet. *UF/IFAS file photo by Lyon Duong.*

Seawalls and other forms of shoreline armoring can have significant harmful effects on coastal ecosystems.[3] Seawalls disrupt natural shoreline processes by preventing sediment from moving naturally along the coast and by reflecting wave energy to adjacent areas. As a result, nearby beaches and marshes may experience increased erosion as they are both deprived of the sediment that would otherwise have naturally replenished them and confronted with overall higher wave energy. Seawalls can also lead to the loss of valuable habitats, such as intertidal zones and shallow coastal waters. The construction of seawalls requires the removal of vegetated shoreline buffer habitats, which then exposes waterways to more direct runoff. Land-based pollutants such as fertilizers, grass clippings, pet waste, and lawn chemicals can flow unimpeded over the edge of the seawall, especially seawalls with turf grass landscapes that are irrigated and mown right up to the seawall. These pollutants further impact marine ecosystems and increase threats to water quality.

Ironically, experts believe that continued reliance on armoring shorelines has actually reduced the resilience of coastal communities because, over time, seawalls often fail to provide adequate protection against rising sea levels and increasingly powerful storms.[4] They may suffer from erosion and structural deterioration due to constant exposure to waves and saltwater, leading to costly maintenance and repairs. When seawalls fail, they can fail catastrophically and lead to significant property loss.[5] Moreover, seawalls can create a false sense of security, encouraging further development in vulnerable areas.[6]

The presence of seawalls exacerbates the consequences of coastal erosion and storm damage, because more structures are put at risk when the seawall eventually fails.[7] As a result, communities that rely heavily on seawalls for protection become less resilient to the impacts of climate change and rising sea levels, putting both human infrastructure and natural ecosystems in jeopardy. In the words of living shoreline researcher Rachel Gittman, in many places across the United States, we have "engineered away our natural defenses." As flooding and erosion risks increase with climate change, restoring and enhancing nature-based coastal protection features will become increasingly important adaptation strategies for coastal communities.

## Living Shorelines: An Alternative to Armoring

Living shorelines are a nature-based approach to coastal erosion control and shoreline protection that uses natural materials and ecological processes to bolster coastal areas. Living shorelines are human-designed constructs that focus on satisfying both human and ecological goals. Unlike traditional hard structures like seawalls and revetments, living shorelines incorporate elements such as saltmarsh plants, mangroves, and oyster reefs to reduce wave energy and promote sediment accumulation.[8] They incorporate natural materials, and their attributes are based on the functions of naturally occurring habitats like mangrove forests, salt marshes, and oyster reefs.

Unlike armored shorelines, living shorelines can provide some of the same benefits of natural shoreline ecosystems, including the following:

- **Reduced erosion**: Living shorelines can help reduce erosion by absorbing wave energy and providing a buffer between the water and the shoreline, while armored shorelines often drive accelerated rates of erosion.

## WHAT IS A LIVING SHORELINE?

"Living shoreline is a broad term that encompasses a range of shoreline stabilization techniques along estuarine coasts, bays, sheltered coastlines, and tributaries. A living shoreline has a footprint that is made up mostly of natural materials. It incorporates vegetation or other living, natural 'soft' elements alone or in combination with some type of harder shoreline structure (e.g., oyster reefs or rock sills) for added stability. Living shorelines maintain continuity of the natural land—water interface and reduce erosion while providing habitat value and enhancing coastal resilience."[a]

Alternate definition 1: "A shoreline management practice that provides erosion control benefits; protects, restores, or enhances natural shoreline habitat; and maintains coastal processes through the strategic placement of plants, stone, sand fill, and other structural organic materials (e.g., biologs, oyster reefs, etc.)."[b]

Alternate definition 2: According to Storm Smart Coasts Florida, "non-structural shore protection measures generally seek to enhance the natural ability of shorelines to absorb and dissipate storm energy without interfering with natural beach, dune, and bank processes."[c]

Beach renourishment, dune stabilization and construction, and stabilizing shorelines with native plants are all examples of non-structural shore protection techniques. Living shorelines are another nature-based shoreline control method, but they are used on low-energy to medium-energy locations rather than on high-energy sites like beaches.

[a] NOAA, *Guidance for Considering the Use of Living Shorelines* (2015), 7, https://www.habitatblueprint.noaa.gov/wp-content/uploads/2018/01/NOAA-Guidance-for-Considering-the-Use-of-Living-Shorelines_2015.pdf.

[b] NOAA Shoreline, "Glossary: Definitions of Common Coastal Terminology," accessed January 31, 2025, https://shoreline.noaa.gov/glossary.html#partj.

[c] Storm Smart Coasts, Florida, "Non-Structural Shore Protection," accessed January 3, 2025, http://fl.stormsmart.org/before/mitigation/non-structural-shore-protection/.

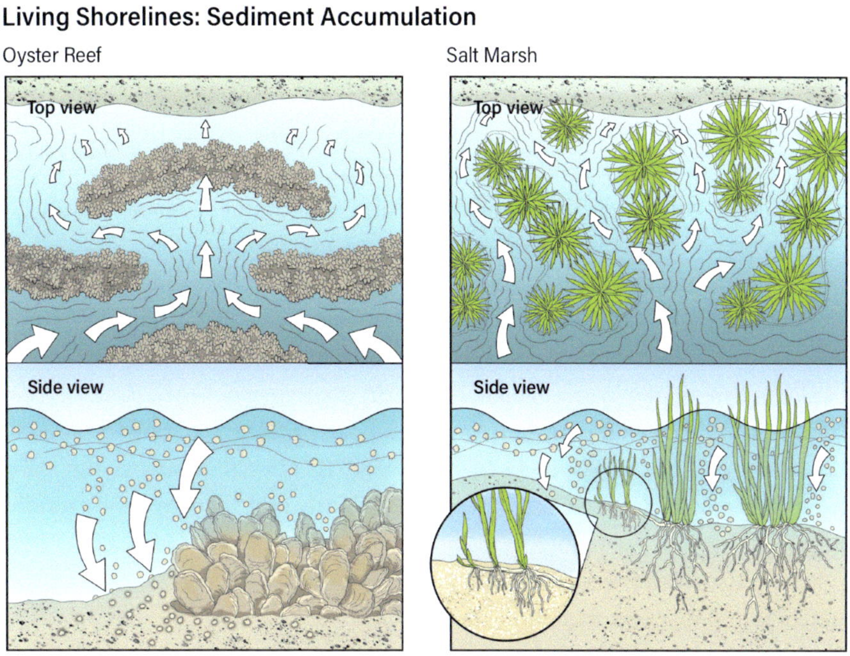


**FIGURE 3.2** How living shorelines can build land. Oyster Reef: The left panel shows an oyster reef from the top and from the side. The reef slows and baffles wave energy through the physical structure of oyster reefs. Calmer water behind the reefs allows sediment to accumulate. Salt Marsh: The right panel shows a salt marsh from the top and from the side. The marsh grasses dissipate the wave energy, and the sediment is dropped in the area. As land builds up, salt marsh plants colonize the new area, and their roots and rhizomes capture additional sediment as shown in the detail. *Illustration by Marjorie Shropshire.*

- **Land accretion**: Accretion is the opposite of erosion and is a process that occurs when there is net sediment capture on a shoreline. By slowing down water flow and baffling wave energy, living shorelines can often capture sediment and build up over time. The ability of the living elements of the shoreline to self-reinforce and gain over time stands in stark contrast to armored shorelines.
- **Enhanced habitat**: Living shorelines can provide habitat for a variety of plants and animals, including fish, birds, and shellfish. Plus, they provide important nursery areas for young fish. In contrast, armored shorelines are often found to support fewer species than natural or living shorelines.[9]
- **Improved water quality**: Living shorelines can help to improve water quality by filtering pollutants and excess nutrients from the water. Also, by slowing down the water flow, sediments are dropped, which makes the water clearer.

- **Enhanced resilience**: Living shorelines can adapt and grow with changing environmental conditions, making them a more sustainable and resilient solution compared to conventional shoreline armoring methods that need to undergo costly retrofits when conditions change.[10]
- **Cost-effectiveness**: Living shorelines cost less to install and, once established, require less ongoing maintenance than traditional hard armoring methods. A recent study of waterfront properties in Florida found that living shorelines were 8.5 times less expensive to install and 2.7 times cheaper to maintain than armored shoreline types.[11]

Living shorelines have gained in popularity, especially in North America, due to their ability to blend with the natural landscape while mitigating the adverse impacts of erosion and sea level rise. In the United States, for example, they have been increasingly implemented along the Atlantic and Gulf coasts to restore degraded shorelines and protect valuable coastal ecosystems.[12] These nature-based solutions also contribute to the reduction of greenhouse gas emissions, as coastal wetlands and marshes are highly efficient carbon sinks.[13] (See the sidebar "Blue Carbon, Peat, and Peatlands," page 21.) These shoreline buffers can also capture and filter nutrients, serving to improve water quality and potentially benefit other aquatic habitats such as seagrasses.[14] By harnessing the regenerative power of nature, living shorelines offer a promising alternative to conventional hard structures, benefiting both human communities and the surrounding coastal environment.

It is crucial to consider factors like sediment dynamics, hydrology, and tidal patterns when designing living shoreline projects to ensure their effectiveness and longevity. (See chapter 4.) The successful implementation of living shorelines depends on careful planning and collaboration between scientists and local communities. Homeowners are a key link in this collaborative relationship, as the vast majority (about 77 percent) of shorelines in Florida are privately owned.[15] Moreover, ongoing monitoring and adaptive designs are essential to assessing the performance of living shorelines and refining their design over time.[16]

Natural shoreline ecosystems are dynamic and allow water, sediment, plants, and animals to move freely between land and water. Although living shorelines maintain this natural connectivity, they do not usually provide the same level of function as natural systems. This is due to the largely urban/suburban (unnatural) context of the uplands and the overall smaller habitat patches that most living shorelines create. The functioning of re-created salt marshes was found to be "significantly impaired" as compared with natural systems, affecting their ability to deliver ecosystem services.[17]

While living shorelines are not considered to be direct replacements for natural marshes and oyster reefs, they can still be considered a "green retrofit" of developed shorelines where natural systems are restored to some extent. If homeowners are fortunate enough to live in an area where the natural shorelines were preserved during development, living shoreline techniques can help to bolster and maintain these natural shorelines. This is especially important if there has been some erosion of these natural areas.

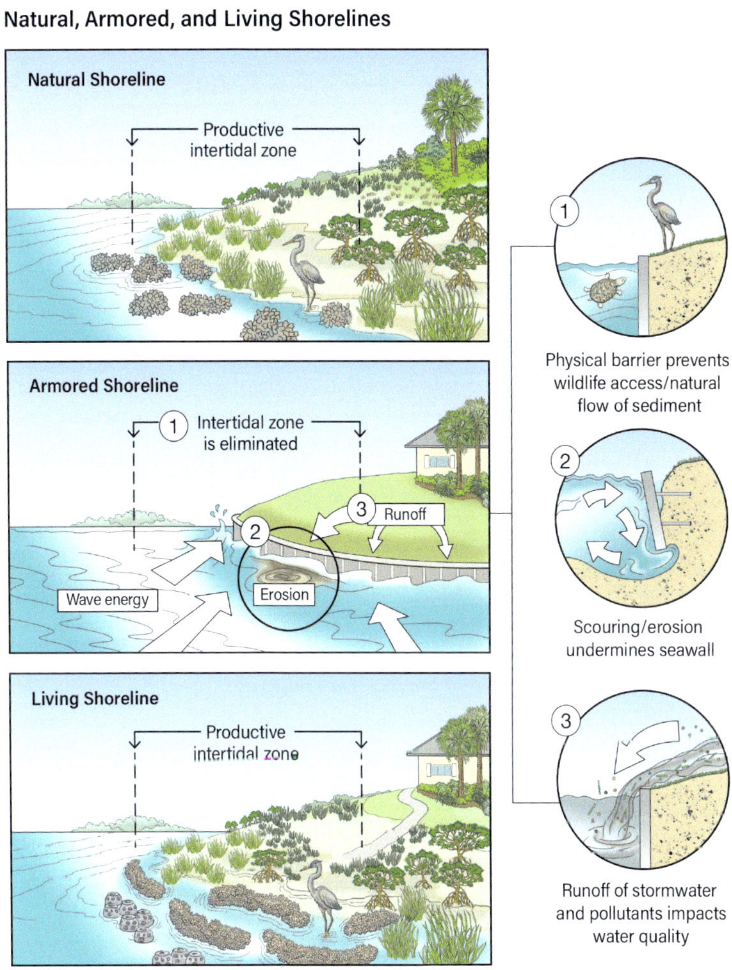


**FIGURE 3.3** Unlike a natural shoreline, armored shorelines break the connection between land and water, effectively eliminating the productive intertidal zone (MLW to MHW) important to life. Armoring also causes problems for wildlife, increases runoff, and is less resilient than a natural shoreline. Living shorelines are conceptually based on natural shorelines, and they help to restore the connection between land and water. *Illustration by Marjorie Shropshire.*

## History of Living Shorelines

The living shoreline field is relatively new and, unfortunately, its early progression is not well documented. Edgar Garbisch (a chemist and founder of the nonprofit Environmental Concern Inc. [https://wetland.org/]) is credited with building the first living shoreline in the mid-1970s.[18] He was inspired after reading *Life and Death of the Salt Marsh* by John and Mildred Teal. Garbisch found success in reestablishing a marsh area by placing rocks along the shoreline of his family home on Maryland's Eastern Shore, filling behind the rocks with sand, and finishing by installing marsh transplants. Later, Garbisch, along with W. W. Woodhouse, Ed Seneca, and Steve Broome, used dredged material in combination with propagation and planting techniques for two marsh grasses—smooth cordgrass (*Spartina alterniflora*) and saltmeadow cordgrass (*S. patens*)—to stabilize a shoreline for the US Army Corps of Engineers. At the time, this approach was called "non-structural erosion control." These efforts are the foundation for the practice of living shorelines as we know it today.[19] In fact, living shoreline practitioners in the Chesapeake Bay credit Garbisch with the creation of the popular marsh sill design that has served as a template for living shorelines in the Chesapeake Bay ever since.[20]

The first scientific study focusing on the concept of living shorelines (non-structural erosion control) was published in 1981.[21] Non-structural erosion control concepts were comprehensively outlined for a general audience in 1992 by a University of North Carolina (UNC) Extension publication encouraging homeowners to take a softer approach to property protection.[22] However, the term living shoreline did not gain popularity until the early 2000s, when it was adopted and popularized by environmental agencies and nonprofits in the US mid-Atlantic region;[23] and the term has since spread throughout North America. The first known use of the term living shoreline in scientific literature was in 2008 and research on living shorelines has increased dramatically in recent years.[24] Overall, this makes living shorelines a relatively young scientific field. But the understanding of coastal ecosystems and the benefits of living shorelines continues to grow. Nature-based solutions such as living shorelines are expected to play a vital role in fostering resilient and sustainable coastal environments worldwide. Florida homeowners and other shoreline managers have been slow to install living shorelines, especially at the private landowner level. Living shorelines are an asset to private property owners and can be managed on smaller scales; they are not just for large public projects, even though the bulk of Florida's living shorelines installed to date have been associated with public lands.

## Alternatives to Armoring

Shoreline armoring, especially the use of vertical seawalls, causes many problems for our coastlines. A study by Rachel Gittman and colleagues titled "Engineering Away Our Natural Defenses" quantified the issue. They found that 12,500 miles (20,116.8 km) or approximately 14 percent of the coastline in the United States is armored. If we keep building seawalls at the present rate of about 124.3 miles (200 km)/year, a third of the coast will be hardened by 2100. That is more than double the amount of present-day armoring![25] Further, Gittman and colleagues found that "over 50% of South Atlantic and Gulf of Mexico coasts are fringed with tidal wetlands that could be threatened by future hardening, based on projected population growth, storm frequency, and an absence of coastal development restrictions." This means that future armoring is likely to come at an especially high cost to our already threatened coastal wetlands in the southeastern United States.

## The Negative Effects of Shoreline Armoring

### Increased Erosion

Shorelines armored with structures like seawalls, revetments, or rock barriers can exacerbate erosion in several ways. While these structures are intended to defend against wave action and protect coastal properties, they disrupt natural coastal processes such as wave transmission and sediment transport, leading to negative consequences for surrounding properties.

- **Reflected Wave Energy**: When waves approach an armored shoreline, they hit the hard surface of the seawall or revetment, and their energy is reflected toward the sea. This reflected wave energy can be stronger than the original wave, leading to increased scouring of the seabed in front of the seawall.[26] As a result, the sediment is eroded away, causing the beach or coastal area in front of the seawall to narrow and recede over time. This can also cause erosion on shorelines adjacent to the structure.
- **Disruptions in Sediment Transport**: Natural coastlines depend on a dynamic balance of sediment transport and deposition, with waves and currents carrying sand and other materials along the shore. Armored shorelines, especially jetties, disrupt this sediment transport process by acting as barriers, preventing the movement of sediment along the coast. This results in a deficit of sand supply to beaches and coastal

environments, which in turn leads to accelerated erosion and loss of shoreline stability.

- **Increased Wave Impact and Scour**: When waves hit a naturally sloped shoreline, they dissipate energy through friction, gravity, and other processes that take place as a wave traverses a gradual slope. When a wave hits an armored shoreline, the waves abruptly encounter and impact the hard surface. This process causes turbulence and erosion (scouring) at the base (toe) of the structure. Over time, this can undermine the foundation of the seawall or revetment and lead to structural failures or eventual collapse.

## Loss of Resilience

While seawalls may offer immediate protection, their long-term impacts on coastal ecosystems and resilience are often detrimental.

- **Erosion and Loss of Beaches**: Seawalls disrupt natural sediment transport along the coast, leading to a loss of sand supply to adjacent shorelines such as beaches, marshes, and dunes. Without a natural source of replenishment, these coastal habitats may narrow or even disappear, reducing the buffer zone that absorbs wave energy during storms. This erosion leaves coastal properties directly exposed to the full force of waves, increasing their vulnerability to storm damage and flooding.
- **Accelerated Coastal Squeeze**: Seawalls restrict the landward migration of coastal ecosystems in response to sea level rise. As sea levels rise due to climate change, natural coastal habitats like salt marshes and mangroves can adapt to the changing water levels by migrating inland. However, seawalls prevent this natural migration, leading to a phenomenon known as "coastal squeeze." This phenomenon can lead to the loss of valuable habitats and the reduction of ecological resilience to climate impacts.
- **Over-Topping and Large Events**: Seawalls may be effective at protecting properties during moderate storms, but during more extreme events, the waves can over-top, or breach, the seawall. Many seawalls were built without consideration for extreme events or how extreme events would increase with climate change and sea level rise. When a seawall is over-topped, this can lead to significant damage to both the seawall structure and the property it was designed to protect. Even if the seawall is not damaged as the flood waters come in, receding flood waters can cause failure of armored shorelines because they are designed to protect

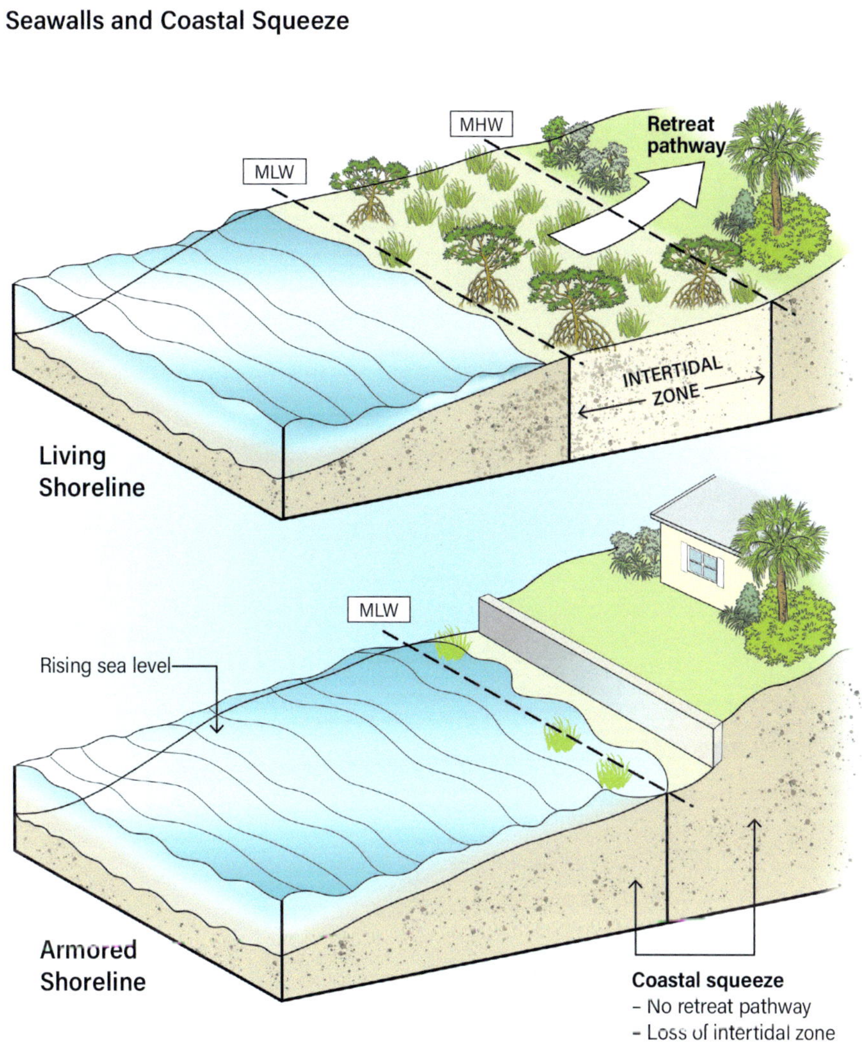


**FIGURE 3.4** Living shorelines provide a retreat pathway for the landward migration of intertidal zone plants and animals. Armored shorelines block the retreat pathway causing coastal squeeze: the loss of intertidal habitat due to rising sea levels and a fixed structure on the shoreline. *Illustration by Marjorie Shropshire.*

**FIGURE 3.5** A major benefit of living shorelines is that they can build up the shoreline and reverse erosion in many cases. These photos are from Bayou Texar in the Florida Panhandle: *(A)* a degraded shoreline consisting of turf grass down to the water's edge. *(B)* The shoreline was enhanced using a system of shell-bag reefs and by the planting of three salt marsh species in the appropriate intertidal zones in December of 2012. *(C)* Five months after installation, in May 2013, the living elements were beginning to coalesce. *(D)* By one year after installation, in December 2013, a significant amount of shoreline area has built up as the salt marsh became fully established and was protected by the adjacent oyster reefs. *Photos by Zachary Schang, Florida Department of Environmental Protection.*

against water only from the seaward side. As water flows back into the waterway, the backside of hard structures can be severely scoured, causing the bank to collapse. Once an armored shoreline is breached, erosion of the exposed bank can occur rapidly, further compromising the stability of the area. By contrast, natural shorelines, such as dunes and marshes, are adapted to handle periodic storm-driven inundation events. The natural slope and vegetation can absorb and dissipate wave energy, while root systems hold tight to sediments both during the onset and recession of floodwaters, providing a more resilient defense against storm events.

- **Costly Maintenance and Adaptation**: As seawalls deteriorate over time due to constant exposure to saltwater and wave action; they require ongoing maintenance and costly repairs. Additionally, as sea levels rise and storms become more intense with climate change, seawalls may need to be built higher or larger to maintain the same level of effectiveness. This ongoing need for retrofit/adaptation can be financially burdensome for communities because it diverts resources away from other important climate resilience and adaptation measures.

## Habitat Loss

Armoring the shoreline severs the connection between land and water. This leads to an unnatural separation of upland and underwater ecosystems, which are otherwise connected by the intertidal zone (the area between high tide and low tide). This can lead to several negative effects on habitat, including the following:

- **Direct Habitat Destruction**: Seawalls are typically constructed by altering the natural shoreline, which often involves some level of excavation and filling to create a stable base for the wall. This process can directly destroy natural coastal habitats in the footprint of the wall construction areas. Habitats such as salt marshes, tidal flats, and dunes—vital for various species of plants and animals—can be damaged during seawall construction. Also, the amplification or reflection of wave energy off the wall's surface can lead to further direct habitat loss by exposing wetlands, oyster reefs, and beaches to a higher and more damaging wave energy.
- **Disruption of Natural Processes**: Armored shorelines interrupt natural processes by altering the flow of sediment along the coast and changing the distribution of sand and other materials. This can result in the degradation or loss of adjacent habitats that depend on sediment supply

for functions such as maintaining elevation. This can lead to erosion or degradation of habitats in front of and adjacent to the seawall.

- **Loss of Biodiversity**: Coastal habitats support a diverse array of plant and animal species, many of which are specially adapted to survive in these unique environments. When seawalls are built, habitat that was previously connected along the shoreline can be fragmented or lost, leading to a decline in biodiversity as species struggle to find suitable places to live and reproduce.[27] In addition, waves that impact vertical seawall structures also resuspend sediments, which can inhibit seagrass growth and threaten the immense biodiversity seagrass meadows support.
- **Loss of Nesting and Feeding Grounds**: Seabirds, marine turtles, horseshoe crabs, and other coastal species often rely on undisturbed coastal habitats for nesting and feeding. Animals such as sea turtles and horseshoe crabs will encounter a hard barrier rather than the gentle slope of a natural shoreline when attempting to move along the intertidal zone for feeding or nesting. One study found that between 12 and 24 percent of potential sea turtle nesting habitat is already obstructed by shoreline

FIGURE 3.6 Seawalls and other forms of armoring threaten critical activities by eliminating or modifying nesting sites and reducing the availability of food sources, impacting the reproductive success and survival of species such as horseshoe crabs (*Limulus polyphemus*) that depend on intertidal shorelines. *Photo by Savanna Barry.*

structures, and these percentages tend to increase after hurricane events.[28] Seawalls threaten critical activities by eliminating or modifying nesting sites and reducing the availability of food sources, impacting the reproductive success and survival of species that depend on intertidal shorelines.[29]

- **Loss of Recreational and Aesthetic Value**: Natural coastal habitats have aesthetic and recreational value for people; they attract tourists and provide opportunities for activities such as birdwatching, hiking, and fishing. When these habitats are lost due to seawall construction, these opportunities diminish. Fewer tourists visiting a community may have a significant economic impact on that community.

## Water Quality Impacts

Many studies have shown that Florida's water quality is getting worse. One driver of this is the loss of shoreline buffer systems that once protected our waters. For example, in the Indian River Lagoon, large declines in natural shorelines, such as mangroves, oysters, and seagrass beds, mean these natural filters are cleaning less water. Meanwhile, increasing amounts of stormwater carrying pollutants and sediments from developed areas, including lawns and streets, flow directly into the Lagoon and other water bodies, largely without being filtered or treated.

- **Habitat Destruction**: Seawalls can lead to the destruction of natural coastal habitats like salt marshes, seagrasses, mangroves, and tidal flats. These habitats act as natural filters, trapping and removing pollutants from the water before they reach the open sea. When these habitats are lost, the water quality deteriorates as the ability to remove pollutants is diminished.
- **Runoff from Impervious Surfaces**: The construction of seawalls is often accompanied by other coastal development, including roads, parking lots, and buildings. These impervious surfaces prevent rainwater from being absorbed into the ground, causing increased runoff. The runoff can carry pollutants from urban areas, such as oil, heavy metals, and fertilizers, into the coastal waters, further degrading water quality.
- **Runoff from Managed Landscapes**: Fertilizers and other chemicals (pesticides, herbicides) applied to managed landscapes such as residential lawns and agricultural fields can flow into water bodies during rain events. Other nutrient sources, such as grass clippings, can lead to unnaturally high nutrient concentrations in the water. High nutrient

concentrations drive algae blooms that can foul the water, decrease oxygen levels in the water, and cause fish kills. Loss of seagrasses allows sediments to become suspended in the water, limiting light penetration and making it harder for seagrass to grow. Armored structures like seawalls make it much easier for this harmful runoff to flow directly into natural water bodies.

## How Living Shorelines Work

Essentially, living shorelines rely on the same processes and functions that make natural shorelines stable. Thus, living shorelines work best on moderate- to low-energy shorelines, just like the natural habitats they mimic. Living shorelines absorb incoming wave energy via shoaling along a gentle slope, the wave baffling effects created by the bending action of plant stems, and the roughness and friction created by the increased complexity. In more moderate wave environments, various submerged reef-like structures or emergent breakwaters can be incorporated into the design to provide an additional layer of wave attenuation, working to further increase roughness and complexity, creating drag on the incoming water and providing an initial shoal that breaks waves before they reach the vegetated zone. Erosion is driven by the energy of waves and the currents that resuspend and export sediment. Thus, living shorelines work by interrupting and dissipating energy, halting and even reversing erosion, and resulting in a stable or gaining (accreting) shoreline system.

Most of the plant species that make up living shorelines are clonal plants that multiply and expand by extending new underground stems called rhizomes. They support each other by contributing to soil oxygenation and gas sharing via rhizome connections, by reducing wave effects on neighbors, and by sharing resources via rhizomatic connections under the soil. In this way, the plants can share the burden of the stressful coastal environment. Their "rhizomatic expansion" is the reason that marsh plants installed as part of a living shoreline coalesce together and become denser over time. It is also the primary mechanism by which marsh vegetation can self-heal after storm damage. Rhizome production is a major contributor to the accumulation of organic material that supports gains in marsh elevation over time.

Oyster reefs, which run parallel along a shoreline, are often key components of living shorelines. Oyster reefs are formed through self-supporting mechanisms: Adult oysters form hard substrates and highly complex surfaces that facilitate the next generation of small oysters settling and surviving to adulthood. Research suggests that the larvae of oysters and other oyster reef inhabitants are drawn to underwater sounds and chemical cues (scents) unique to oyster reefs.[30] They are like beacons calling out "settle here, it's safe." Living shorelines leverage these and other amazing functions of natural shoreline ecosystems. In fact, a study at a living shoreline in Cedar Key, Florida, found that the underwater soundscape was significantly enhanced at the site, likely leading to the further attraction of larvae to the area.[31]

## Types of Living Shorelines

Types of living shorelines are listed here for reference and are further defined in chapter 4.

### Nature-Based Living Shorelines

Suitable for protected locations with gradual slopes and small waves, nature-based living shorelines fall on the "greener" end of the spectrum. They incorporate natural elements like mangroves, smooth cordgrass, and oyster reefs. Depending on the site, just adding plants may be enough to decrease erosion. As plants and oysters grow, the shoreline becomes more stable and more resilient to sea level rise and tidal flooding. In addition, sediment may collect on the shoreward side of these living building blocks, essentially growing the shore. Nature-based systems are the easiest and least expensive to install, and unlike seawalls, they strengthen over time.

### Hybrid Living Shorelines

Hybrid systems combine natural elements with hard infrastructure and are suitable for sites with medium to high energy and a less gradual slope. They may be used to retrofit existing seawalls to increase habitat, to shore up a failing structure and extend its lifespan, or for aesthetic reasons. A common structural element is coquina riprap, which is natural to Florida and provides habitat for plants and animals, including oysters. Riprap can be placed inshore of oyster reefs and intertidal plants, or it can be added to the front of an existing seawall to decrease wave-driven scour that can undermine the structure's base.

Hybrid systems are often installed on canal front properties that have limited lateral space, including those with seawalls. Coquina stone can be placed at the base of a seawall, or it can be used to construct mangrove planters that are decorative and functional. A hybrid system might also incorporate wall inserts or

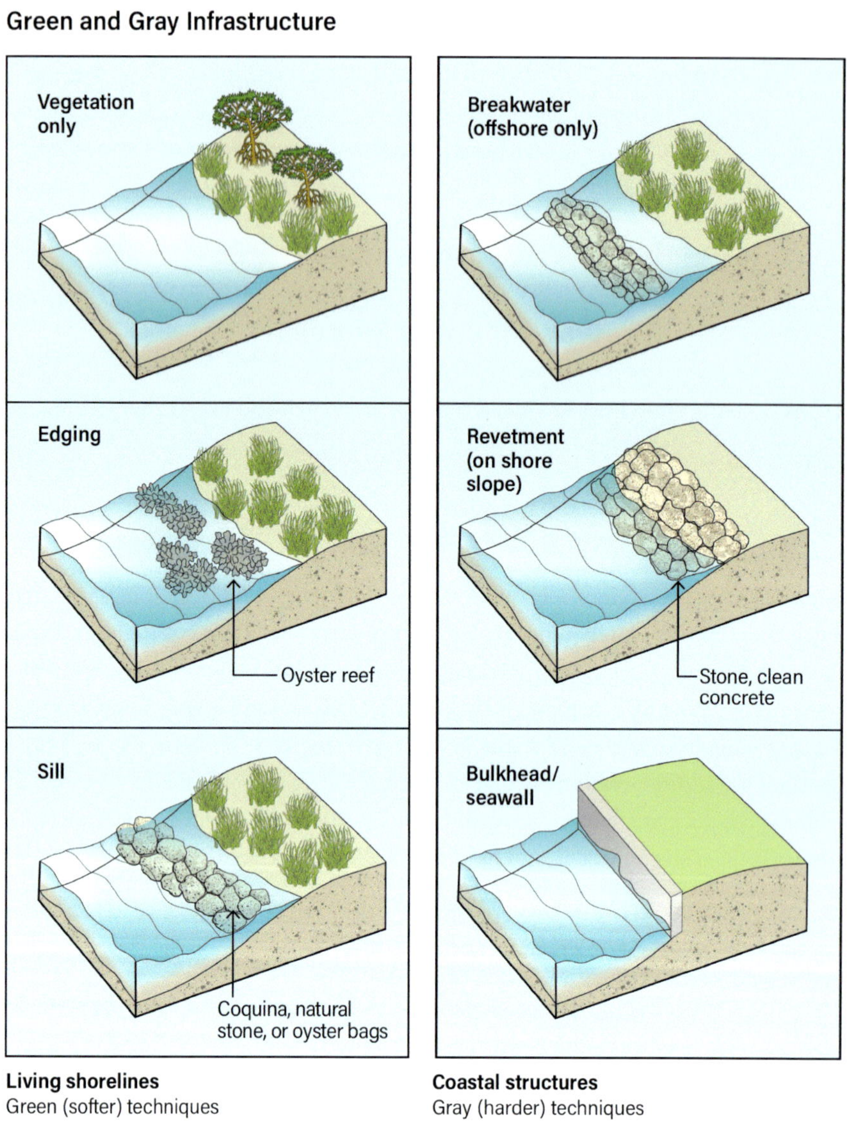


**FIGURE 3.7** Examples of green (softer) and gray (harder) infrastructure. Wherever possible, greener solutions should be used. *Illustration by Marjorie Shropshire. Adapted from NOAA, Guidance for Considering the Use of Living Shorelines (2015), 8, https://www.habitatblueprint.noaa.gov/wp-content/uploads/2018/01/NOAA-Guidance-for-Considering-the-Use-of-Living-Shorelines_2015.pdf.*

sculpted panels attached to the face of the seawall. These increase surface area for fish and invertebrates, can retrofit a barren canal front shoreline into a refuge for small creatures, and can be aesthetically pleasing.

## Ecosystem Services of Living Shorelines: Local to Global

From carbon sequestration to human health and well-being, living shorelines provide many ecosystem services to people and the communities they live in. At the global level, they store and sequester carbon, becoming part of blue carbon capture performed by coastal and ocean systems, as discussed in chapter 1. Thus, local living shorelines influence Florida as a whole, and beyond, since we all share the same atmosphere. At the local level, living shorelines improve human health and well-being. More habitat leads to improved fisheries, which create more food and jobs. Reduced wave energy and accrued sediment protect people and structures from erosion, flooding, and storm events. Better water quality allows for the enjoyment of recreational activities like swimming, fishing and boating, and helps maintain income from tourism. These benefits are important to Florida's economy.

## Living Shorelines Benefit Both People and Nature

Living shorelines are defined as "any shoreline management system that is designed to protect or restore natural shoreline ecosystems through the use of natural elements and, if appropriate, man-made elements."[32] The elements and benefits of living shorelines are based on naturally occurring ecosystems. In coastal Florida, these ecosystems are principally salt marshes, mangrove forests, and oyster reefs. Natural systems are dynamic and allow water, sediment, plants, and animals to move freely, unlike seawalls, which create static, vertical barriers to movement. Living shorelines are designed to maintain this natural connectivity. Although living shorelines do not provide the same level of function as natural systems, they can still provide a range of benefits to people and nature. They are also not restricted to coasts. Living shorelines on creeks, rivers, lakes, and even on Florida's 76,000 stormwater ponds can reduce erosion and improve habitat. (Nature-based shoreline enhancements are sometimes called "riparian buffers" in freshwater systems.) Retrofitting stormwater ponds with edge vegetation also helps to absorb chemical pollutants running off the land, so the water quality is improved.[33]

**FIGURE 3.8** Florida living shorelines support resilient communities. Salt marshes provide valuable services, such as (*1*) capturing carbon, (*2*) trapping sediment, growing new biomass, and increasing surface elevation. (*3*) Living shorelines dampen wave energy and reduce erosion. Hardened shorelines (*4*) negatively impact animals, and are (*5*) subject to erosion, causing failure during storm surge events. *Illustration by Marjorie Shropshire. Adapted from NOAA Habitat Blueprint, "Living Shorelines Support Resilient Communities," accessed April 9, 2025, https://www.habitatblueprint.noaa.gov/living-shorelines/*

**Circle 1** shows carbon sequestration by salt marshes. One acre of salt marsh provides several ecosystem services, each valued between $981 and $6,471 per year. As living shorelines establish, their benefits typically increase over time, in contrast to armored shorelines that have diminishing benefits over time.

**Circle 2** shows the processes by which salt marshes—a main component of many living shoreline designs—trap sediments and organic matter and grow new biomass, potentially leading to increased surface elevation.

**Circle 3** illustrates how much living shorelines can dampen wave energy and reduce erosion. Even narrow fringing marshes like those created in many living shoreline designs have been shown to slow waves and reduce shoreline erosion. One study showed that densely growing cordgrass (*Spartina* spp.) in a salt marsh can dissipate wave energy by up to 50 percent within the first 16 feet (5 m) and 95 percent after 98 feet (30 m). Another study found that 330 feet (100 m) of mangroves can reduce wave heights by up to 60 percent.

**Circle 4** illustrates how the cumulative impacts of hardened shorelines manifest at a system level. Armoring of a few small sections of shoreline may have only local adverse impacts, but as more shorelines become armored, systems can reach thresholds where ecosystem services begin to be noticeably degraded. For example, armored shorelines can negatively affect animals that burrow in the sediment, like clams and worms (called infauna) in adjacent areas. Infauna are important food sources for many consumers and, in turn, predators in coastal food webs, including popular sport fish like red drum and snook. Therefore, lower infauna densities adjacent to seawalls can lead to diminished predator densities and less productivity system wide.

*(Fig. 3.8 caption continued)* **Circle 5** refers to the fact that during severe storms bulkheads can fail. Findings from a study along North Carolina's sheltered coasts that found marshes (with and without sills) outperformed bulkheads during Category 1 Hurricane Irene in 2011. The marsh and sill configurations accreted sediment, while 75 percent of regional bulkheads surveyed were damaged. In another example, 58 percent of bulkheads were destroyed during Hurricane Hugo in 1989 along South Carolina's coast. More recently, widespread seawall failure has occurred in Punta Gorda, Florida, after Hurricane Irma in 2017 and again after Hurricane Ian in 2022. In fact, while crews were still working to repair seawalls damaged by Hurricane Irma, more than seven additional miles (11.3 km) of seawalls failed in Punta Gorda after Hurricane Ian.

Living shorelines are not considered true restoration projects, which focus only on ecological goals. Instead, they combine human use and benefits with ecological goals. There is no single solution that can be used when planning and building a living shoreline since each site is unique. Some people have observed that living shorelines are "a mix of both art and science. They are always a balancing act that strives to provide the dynamic equilibrium between natural forces and structures that can deliver both shoreline erosion protection and increased habitat values. This dynamic equilibrium allows a living shoreline to be able to shape and reshape itself in response to external processes yet remain stable enough to provide long-term erosion control and habitat enhancement benefits to a reach of shoreline."[34]

The concept of living shorelines is based on the premise that human-created systems will fulfill some of the functions of natural systems, including coastal protection. According to Rodriguez-Calderon, "living shorelines were estimated to be able to enhance ecosystem services capacity by more than 90% compared to their previous status."[35] Clearly, they are valuable alternatives to armoring, especially in developed areas. Armored shorelines and sea level rise increase flooding and erosion. Living shorelines reduce the energy of floodwaters and protect the coast from storms because of the ability of salt marshes and oyster reefs to capture and retain sediment.

Living shorelines are dynamic and may keep pace with sea level rise if their constructed salt marshes and oyster reefs have enough available sediment. However, their lifespan may be limited by coastal squeeze and the loss of intertidal habitat resulting from rising seas combined with fixed anthropogenic modifications such as buildings or armoring. But even if lateral space is limited, living shorelines are still valuable for their ability to provide ecosystem services and reduce habitat loss. In areas where landward migration of marshes and wetlands is possible, living shorelines will be more successful in the long term, and they are "one of the only methods that will be able to both adjust to future climate conditions and preserve essential ecological functions."[36]

## BENEFITS OF LIVING SHORELINES

### STRUCTURAL BENEFITS

- Reduce wave energy
- Prevent erosion of land
- Resist storm impacts
- Preserve or enhance resiliency
- Coastal protection for people and property

### ECOLOGICAL BENEFITS

- Improve water clarity
- Carbon sequestration
- Remove nutrients that fuel algae blooms
- Support a natural connection between land and water
- Preserve wildlife habitat
- Habitat for fish, crabs, and other marine life

### BEAUTY AND USE BENEFITS

- Provide aesthetic value
- Create easier access to the waterfront for kayaking and other activities
- Help connect homeowners to the natural world
- Attract birds and other wildlife

### FINANCIAL BENEFITS

- Lower up-front cost helps to save money
- Are self-sustaining over time, unlike seawalls
- Require less expensive maintenance
- Support tourism and recreation-based economies. Ocean-related tourism and recreation in the United States generates $89 billion in economic output and $9.1 million in jobs (2010).[a]

### ADAPTATION TO SEA LEVEL RISE

- Assist in landward migration
- Trap sediments
- Capture sediment to prolong persistence of created marshes; particularly when oyster reefs are part of the living shoreline
- Can self-repair after storms because of the living elements of the shoreline
- Even where landward migration is restricted by infrastructure behind the living shoreline, the benefits of maximizing ecosystem services may still be substantial.

[a] Judith T. Kildow, Charles S. Colgan, Jason D. Scorse, Pat Johnston, and Matt Nichols, "State of the US Ocean and Coastal Economies 2014," Publication 1 (National Ocean Economics Program, 2014), http://cbe.miis.edu/noep_publications/1/.

## Determining Whether a Living Shoreline Is Right for a Property

In a 2019 survey conducted by the University of Florida, waterfront property owners were asked to rate their overall satisfaction with their shoreline type (armored, hybrid, living shoreline, or natural). The majority of respondents reported high levels of satisfaction with their shoreline, no matter what type they had, but property owners with natural and hybrid shorelines reported the highest rates of satisfaction.[37] For their shoreline type, respondents were asked to rate access, recreation, aesthetics, maintenance, protection, and environmental benefits. Hybrid shorelines received the highest scores in the areas of access, recreation, and maintenance. Property owners thought armored shorelines provided the most protection. Natural shorelines were rated highest for aesthetics, and living shorelines received the highest score for environmental benefits. Overall, these results indicate that the right shoreline type for a property might vary depending on what type of shoreline service is most important to the homeowner or property manager. However, in general, natural or hybrid shorelines appear to lead to slightly higher overall satisfaction. Homeowners with a stable or gaining shoreline of any type had significantly higher satisfaction than homeowners with an eroding shoreline. If a living shoreline is installed and established to the point that it mimics a natural shoreline (the category with the highest satisfaction) and if the shoreline is stable or gains material, this research predicts a happy homeowner would result.

In the same survey, participants were asked about their considerations when making shoreline management decisions. Effectiveness was identified as the most important factor in shoreline management decision-making, while durability and cost rated second and third, respectively.[38] Participants perceived armored shorelines to be the most effective. However, more natural shoreline types ranked higher in all other aspects. Overall, these results suggest that homeowners tend to generally be aware of at least some of the drawbacks of armoring their shoreline. Data also suggests that many are willing to sacrifice some shoreline functions in order to obtain what they perceive as the most effective erosion-prevention option. But a properly designed living shoreline can be just as or more effective than a seawall while avoiding the negative impacts of armoring. Further, survey results showed that living shorelines were 8.5 times cheaper to install and 2.7 times cheaper to maintain than armored shorelines.[39] The purpose of this book is to help people make the most informed decision about what is right for their particular shoreline.

## Living Shoreline Longevity

Florida living shorelines pioneer Tom Ries said it best: "A seawall will hold the line for a finite amount of time while living shorelines can advance the line for a potentially indefinite amount of time." This statement is based on industry estimates,[40] which state that the average longevity of a seawall is 20 years (or less) for a poorly built or poorly maintained seawall, while the best-case scenario is a lifespan of 30 to 50 years for a properly built or maintained seawall. This is not that long, considering the typical length of a home loan is 30 years. By contrast, a properly designed living shoreline will not need to be replaced, only maintained. Such maintenance might entail, for example, replacing plants that die in the early years or restacking oyster bags after storms.

The bottom line is that there is no reason to expect that a well-designed living shoreline would ever need to be replaced, while every seawall will eventually need to be replaced.

## Common Homeowner Questions About Living Shorelines

### Will It Work?

Despite all the evidence covered in this chapter, some people might still be wondering if living shorelines really work. Our human nature leads us to put more trust in infrastructure like walls than in natural elements like plants and oysters. How do we know the living shorelines approach will work? There are some great examples of living shorelines outperforming seawalls and bulkheads. A paper published by Gittman and colleagues in 2014 documented a case study of shorelines in North Carolina that received a direct hit from a Category 1 hurricane.[41] The study presented images of two shorelines that were only about 1,600 feet (500 m) apart from each other and that therefore encountered the same conditions during the storm. The living shoreline weathered the storm with minimal impacts (and actually captured sand), while the bulkhead suffered catastrophic damage. An article available from *Scientific American* describes this case study in more detail.[42]

Another example from Florida involves a large-scale living shoreline project called Project Greenshores in Pensacola. The project involved offshore breakwaters that created a huge gain in marsh habitat behind the structures. There were improvements in seagrass habitat, fish and bird use, and many other measures of success. A major hurricane, Hurricane Ivan, struck Pensacola in 2004, just two years after this project was installed. The project sustained some damage but recovered quickly and with minimal intervention, and it still maintains a high level of function as of this writing. Project Greenshores was also hit by Hurricane Sally in 2020, another major test that it passed with flying colors. Both Hurricanes Ivan and Sally led to catastrophic failure of seawalls at nearby properties such as Sander's Beach and Camp Timpoochee.

## What About Maintenance?

When thinking about maintenance, it may be initially attractive to think about seawalls as a "build it and forget it" solution. But the fact is that seawalls are the strongest they will ever be at the beginning of their lifespan and will weaken over time. And they will incur compounding costs over time as the wall degrades, making the property more vulnerable to additional impacts and additional costs. Even seawalls that are not hit by a hurricane will eventually need expensive maintenance and repair or replacement, and these needs will increase as the seawall ages and as sea level rise progresses.

By contrast, living shorelines incur compounding gains because they are less expensive to install initially, and over time, living shorelines become self-sustaining. Living shorelines are the weakest right at the beginning, when new plants are not fully rooted and new reefs have low oyster cover. As long as the living shoreline has been designed and sited correctly, there will be an initial period of maintenance (such as replacing plants that do not survive) followed by a potentially indefinite period of low maintenance (such as mangrove management; see below) or no maintenance. Further, living shorelines can even self-repair after mild to moderate storm damage.

Initial maintenance efforts for nature-based living shorelines include replanting in underperforming areas to ensure coalescence of vegetation. Oyster reef substrates should be inspected soon after installation to make sure they did not move, and some replacement should be anticipated. Over time, the maintenance reduces, but an average of 10 to 15 percent of the initial installation costs should be anticipated for the maintenance phase (the first three to five years after installation).

By comparison, armored shorelines initially require very little maintenance, but over time (from year two onward) repairs to address cracking and settling issues will be needed. Property owners should budget 100 percent of the initial installation cost (e.g., $15,000 to $60,000 for a 100-foot [30.5 m] seawall[43]) for inspections, repairs, and maintenance over the 30- to 50-year lifespan of the structure. According to 2004 estimates from Coastal Systems International, repairs may cost $100 to $400 per linear foot of wall. Retaining wall replacement, depending

**FIGURE 3.9** Examples of how seawalls fail during hurricanes. In contrast to living and natural shorelines, widespread and catastrophic failures of seawalls and bulkheads are common after hurricanes. An early example of this was documented in 2011 after Hurricane Irene hit North Carolina. Photo *(A)* shows a bulkhead structure before the storm (May 2011), while photo *(B)* shows the same bulkhead in October 2011, after the storm. The bulkhead required extensive (and expensive) repairs, including complete replacement and backfill using 22 truckloads of sand. Photos *(C)* and *(D)* show a neighboring living shoreline that sustained no damage. Photo *(E)* shows seawall failures in Camp Timpoochee, reported in the Pensacola area after Hurricane Sally in 2020. Photo *(F)* shows similar damage from Punta Gorda, in southwest Florida, after Hurricane Ian in 2022. In both cases the repairs were costly for property owners to address. *Photos A, B, C, and D by Dr. Rachel Gittman. Photo E by Savanna Barry. Photo F by the City of Punta Gorda, Florida.*

**FIGURE 3.10** Project Greenshores, a large living shoreline in Pensacola, Florida, was installed in 2002 and remains successful despite weathering impacts from hurricanes. These photos are of approximately the same area where you can see the roadside marsh and the oyster reefs outside the barrier islands that were built as part of this project. Photo *(A)* was taken in December 2019, when much of the vegetation was browned for the winter. Photo *(B)* was taken after Hurricane Sally in September 2020; you can see sand washed up on the shoreline marsh area next to the road; Photo *(C)* was taken in August 2024, when the roadside marsh and the islands are green with vegetation. See Dauphin Island Sea Lab for more information on this project. *Photos by Darryl Boudreau, Boudreau Photography.*

**FIGURE 3.11** Examples of living shorelines performing well in hurricanes. These photos show living shoreline projects in Florida before and after various hurricane and tropical storm events. The Needlerush Point living shoreline on Perdido Key *(A)* before and *(B)* after Hurricane Sally in 2020. The living shoreline at La Serena condominiums in Pensacola *(C)* before and *(D)* after Hurricane Sally. The living shoreline retrofit at Joe Rains Beach in Cedar Key *(E)* before and *(F)* after Hurricane Michael in 2018. The South Cove Natural Area living shoreline in West Palm Beach on September 6th *(G)* and September 15th *(H)*, before and after Hurricane Irma in 2017. *Photos A and B by Rick O'Connor. Photos C and D by Charles Krupnick. Photos E and F by Savanna Barry. Photos G and H by Palm Beach County Department of Environmental Resources Management.*

upon height, anchoring and associated drainage components, may cost $500 to $1,000 per linear foot (0.3 m).[44]

## What About the View?

If the property is in a region of Florida where mangroves are prevalent, it is likely that mangroves will become the dominant vegetation type. Florida's 1996 Mangrove Trimming and Preservation Act (MTPA) set out laws protecting mangroves and regulating how they may be trimmed and managed. Mangroves provide an excellent wind break in addition to their wave attenuation benefits. To maintain water views, mangroves can be trimmed using a windowing approach so that the property owner can still retain the protective benefits of the taller tree canopy. If the property owners are concerned about taller mangroves affecting their views even with the window trim, in some cases they may opt to trim and maintain mangroves at a height of as low as 6 feet, provided they do it in a way that complies with the MTPA. For example, mangroves that have reached a height of 10 feet (3 m) or taller must initially be trimmed by a certified professional arborist or mangrove trimmer before the homeowner can resume self-management of the mangroves. Often, shoreline property owners and managers can add provisions for mangrove trimming and maintenance in their initial living shoreline construction permit so there will not be any extra steps related to mangrove trimming. Consult the Florida Department of Environmental Protection for more information about mangrove trimming and maintenance.

## What Will It Cost?

Installation costs for shoreline projects vary widely depending on factors such as material choice, existing site conditions, site access, permitting, and whether the property owners or managers work with a team of professionals. With this caveat, here are some rough estimates of up-front installation costs (and remember to also factor in maintenance costs covered above):

- Costs to install a nature-based living shoreline range between $15 to $55 per linear foot (0.3 m). Living shorelines that incorporate more substantial reef structures in addition to plants could cost between $100 and $400 per linear foot on average, and up to $1,200 per linear foot if the largest and most expensive breakwater materials are chosen.
- For installing armored shorelines, riprap revetments may cost between $600 and $1,800 per linear foot (0.3 m). Here the cost depends mostly

**FIGURE 3.12** Photos of mangrove trimming options: *(A)* shows a window trim with selected mangroves growing tall; *(B)* shows a hedge trim where all the mangroves are trimmed to a certain height; *(C)* shows a natural (no trim) state. *Photo A, Proper Mangrove Windowing, Indian Rocks Beach ICWW / Thomas Ries—Ecosphere Restoration Institute, 2024. Photo B, City of Naples, Florida. Photo C, UF/IFAS file photo by Tyler Jones.*

upon the size (diameter) of rocks installed. Bulkheads/seawalls range from $300 to $1,100 per linear foot, and retaining walls range from $1,500 to $3,500 per linear foot depending largely on the wall material, number of anchors, the cap size, and the existing substrate type at the site.

## Learn More

The following chapters will cover planning, permitting, building, and maintaining a living shoreline and upland buffer habitats on shoreline properties. Even if homeowners or other shoreline managers plan to work with a team of professionals, this book provides important guidelines to help them make informed decisions and know what to ask for from contractors or consultant teams. It also provides information on ongoing maintenance and other shoreline friendly actions that can be implemented after the living shoreline is installed and established.

## Notes

1. Diane Mitsova, Chris Bergh, and Greg Guannel, *Suitability Analysis for Living Shorelines Development in Southeast Florida's Estuarine Systems* (Florida Atlantic University, April 2016), https://maps.coastalresilience.org/seflorida/methods/Living_Shorelines_Final_Report_05_06_16.pdf.
2. Ehud Galili, Jonathan Benjamin, Vered Eshed, Baruch Rosen, John McCarthy, and Liora Kolska Horwitz, "A Submerged 7000-Year-Old Village and Seawall Demonstrate Earliest Known Coastal Defense Against Sea-Level Rise," *PLOS One* 14, no. 12 (2019): e0222560.
3. J. E. Dugan, K. A. Emery, M. Alber, et al., "Generalizing Ecological Effects of Shoreline Armoring Across Soft Sediment Environments," *Estuaries and Coasts* 41, Suppl. 1 (2018): 180–96, https://doi.org/10.1007/s12237-017-0254-x.
4. Rachel K. Gittman, F. Joel Fodrie, Alyssa M. Popowich, et al., "Engineering Away Our Natural Defenses: An Analysis of Shoreline Hardening in the US," *Frontiers in Ecology and the Environment* 13, no. 6 (2015): 301–7; John N. Kittinger and Adam L. Ayers, "Shoreline Armoring, Risk Management, and Coastal Resilience Under Rising Seas," *Coastal Management* 38, no. 6 (2010): 634–53.
5. Vitaly Feygin, "Flexible Seawalls: Modified Bulkhead Approach," *Practice Periodical on Structural Design and Construction* 17, no. 2 (2012): 74–80; Mantripathi P. R. Jayaratne, Buddhika Premaratne, Abimbola Adewale, et al., "Failure Mechanisms and Local Scour at Coastal Structures Induced by Tsunami," *Coastal Engineering Journal* 58, no. 4 (216): 1640017, https://doi.org/10.1142/S0578563416400179; Keith Adams and Muhammad Heidarzadeh, "Extratropical Cyclone Damage to the Seawall in Dawlish, UK: Eyewitness

Accounts, Sea Level Analysis and Numerical Modelling," *Natural Hazards* 116 (2023): 637–62, https://doi.org/10.1007/s11069-022-05692-2; Roshanak Nateghi, Jeremy D. Bricker, Seth D. Guikema, and Akane Bessho, "Statistical Analysis of the Effectiveness of Seawalls and Coastal Forests in Mitigating Tsunami Impacts in Iwate and Miyagi Prefectures," *PLOS One* 11, no. 8 (2016): e0158375.

6. Nateghi et al., "Statistical Analysis of the Effectiveness of Seawalls."
7. Kittinger and Ayers, "Shoreline Armoring, Risk Management."
8. NOAA Habitat Blueprint, "Living Shorelines," accessed February 1, 2025, https://www.habitatblueprint.noaa.gov/living-shorelines/.
9. Rachel K. Gittman, Steven B. Scyphers, Carter S. Smith, Isabelle P. Neylan, and Jonathan H. Grabowski, "Ecological Consequences of Shoreline Hardening: A Meta-Analysis," *BioScience* 66, no. 9 (2016): 763–73.
10. Restore America's Estuaries, *Living Shorelines: From Barriers to Opportunities* (Arlington, VA, 2015), https://estuaries.org/wp-content/uploads/2019/02/Living-Shorelines-From-Barriers-to-Opportunities.pdf.
11. Savanna Barry, Laura K. Reynolds, Anna E. Braswell, Rachel K. Gittman, Steven B. Scyphers, and Ashley R. Smyth, "Perceived Effectiveness Drives Shoreline Decision-Making for Florida's Waterfront Property Owners," *Ocean and Coastal Management* 258 (2024): 107353, https://doi.org/10.1016/j.ocecoaman.2024.107353.
12. Ariana E. Sutton-Grier, Rachel K. Gittman, Katie K. Arkema, et al., "Investing in Natural and Nature-Based Infrastructure: Building Better Along Our Coasts," *Sustainability* 10, no. 2 (2018): 523, https://doi.org/10.3390/su10020523.
13. Amanda M. Nahlik, and M. Siobhan Fennessy, "Carbon Storage in US Wetlands," *Nature Communications* 7, no. 1 (2016): 1–9; U.S. Environmental Protection Agency (EPA), "Coastal Resiliency," accessed October 1, 2023, https://www.epa.gov/green-infrastructure/coastal-resiliency.
14. Cindy M. Palinkas, Miles C. Bolton, and Lorie W. Staver, "Long-Term Performance and Impacts of Living Shorelines in Mesohaline Chesapeake Bay," *Ecological Engineering* 190 (2023): 106944, https://doi.org/10.1016/j.ecoleng.2023.106944.
15. Pamela Pogue and Virginia Lee, "Providing Public Access to the Shore: The Role of Coastal Zone Management Programs," *Coastal Management* 27 (1999): 219–37.
16. Margaret A. Myszewski, and Merryl Alber, *Living Shorelines in the Southeast: Research and Data Gaps.* Georgia Coastal Research Council, University of Georgia, Athens, GA, 2016.; Molly Mitchell, and Donna Marie Bilkovic, "Embracing Dynamic Design for Climate-Resilient Living Shorelines, *Journal of Applied Ecology* 56, no. 5 (2019): 1099–105, https://doi.org/10.1111/1365-2664.13371; Carter S. Smith, Morgan E. Rudd, Rachel K. Gittman, et al., "Coming to Terms with Living Shorelines: A Scoping Review of Novel Restoration Strategies for Shoreline Protection," *Frontiers in Marine Science* 7 (2020): 434.
17. K. L. Spencer and G. L. Harvey, "Understanding System Disturbance and Ecosystem Services in Salt Marshes: Integrating Physical and Biogeochemical Processes," *Estuarine, Coastal and Shelf Science* 106 (2012): 23–32.

18. Gabriel Popkin, "Reshaping the Chesapeake Bay, One Living Shoreline at a Time," *Washington Post,* March 14, 2016, https://www.washingtonpost.com/national/health-science/reshaping-the-chesapeake-bay-one-living-shoreline-at-a-time/2016/03/14/9c223a4c-c51d-11e5-8965-0607e0e265ce_story.html.
19. William Walton Woodhouse Jr., Ernest D. Seneca, and Stephen W. Broome, *Propagation of "Spartina alterniflora" for Substrate Stabilization and Salt Marsh Development,* Technical Memorandum No. 46 (US Army Corps of Engineers, Coastal Engineering Research Center, Fort Belvoir, VA, 1976), https://repository.library.noaa.gov/view/noaa/46908; E. W. Garbisch Jr., P. B. Woller, and R. J. McCallum, *Salt Marsh Establishment and Development,* Technical Memorandum 52 (U.S. Army Corps of Engineers, Coastal Engineering Research Center, Fort Belvoir, VA, 1975), https://usace.contentdm.oclc.org/digital/collection/p266001c0111/id/7016/.
20. Dave Wheelan, "The Late Ed Garbisch and His Legacy of Environmental Concern for Wetland Protection," *Chestertown Spy,* August 20, 2018, https://chestertownspy.org/2018/08/20/the-late-ed-garbisch-and-his-legacy-of-environmental-concern-for-wetland-protection/; Zach Royle, *Living Shorelines* (American Littoral Society, New Jersey), accessed February 1, 2025, https://www.cleanoceanaction.org/fileadmin/editor_group2/Water_Quality/Living_Shorelines_ALS.pdf.
21. Smith et al., "Coming to Terms with Living Shorelines."
22. Stephen W. Broome, Spencer M. Rogers Jr., and Ernest D. Seneca, *Shoreline Erosion Control Using Marsh Vegetation and Low-Cost Structures,* UNC-SG-92–12 (North Carolina Sea Grant, University of North Carolina, 1992).
23. C. A. Currin, W. S. Chappell, and A. Deaton, "Developing Alternative Shoreline Armoring Strategies: The Living Shoreline Approach in North Carolina," in *Puget Sound Shorelines and the Impacts of Armoring—Proceedings of a State of the Science Workshop, May 2009,* edited by H. Shipman, M. N. Dethier, G. Gelfenbaum, K. L. Fresh, and R. S. Dinicola (US Geological Survey Scientific Investigations Report 2010–5254, 2010), https://pubs.usgs.gov/sir/2010/5254/.
24. Smith et al., "Coming to Terms with Living Shorelines."
25. Gittman, et al., "Engineering Away Our Natural Defenses."
26. Jonathon R. Miles, Paul E. Russell, and David A. Huntley, "Field Measurements of Sediment Dynamics in Front of a Seawall," *Journal of Coastal Research* 17, no. 11 (2001): 195–206, http://www.jstor.org/stable/4300163.
27. M. G. Chapman, "Paucity of Mobile Species on Constructed Seawalls: Effects of Urbanization on Biodiversity," *Marine Ecology Progress Series* 264 (2003): 21–29.
28. Blair Witherington, Shigetomo Hirama, and Andrea Mosier, "Barriers to Sea Turtle Nesting on Florida (United States) Beaches: Linear Extent and Changes Following Storms," *Journal of Coastal Research* 27, no. 3 (2011): 450–58.
29. H. Galbraith, R. Jones, R. Park, et al., "Global Climate Change and Sea Level Rise: Potential Losses of Intertidal Habitat for Shorebirds," *Waterbirds* 25, no. 2 (2002): 173–83; Sarah

E. Hirsch, Madison Toonder, Jennifer D. Reilly, Shelby R. Hoover, and Justin R. Perrault, "Responses of Three Nesting Sea Turtle Species to Hard-Armoring Structures," *Frontiers in Marine Science* 9 (2022): 980715, https://doi.org/10.3389/fmars.2022.980715.

30. Mario N. Tamburri, Christopher M. Finelli, David S. Wethey, and Richard K. Zimmer-Faust, "Chemical Induction of Larval Settlement Behavior in Flow," *Biological Bulletin* 191, no. 3 (1996): 367–73; Ashlee Lillis, D. B. Eggleston, and D. R. Bohnenstiehl, "Estuarine Soundscapes: Distinct Acoustic Characteristics of Oyster Reefs Compared to Soft-Bottom Habitats," *Marine Ecology Progress Series* 505 (2014): 1–17; Ashlee Lillis, DelWayne R. Bohnenstiehl, and David B. Eggleston, "Soundscape Manipulation Enhances Larval Recruitment of a Reef-Building Mollusk," *PeerJ* 3 (2015): e999.
31. Audrey Looby, Laura K. Reynolds, Ashley M. McDonald, Savanna C. Barry, Mark Clark, and Charles W. Martin, "Intertidal Soundscapes of Hardened and Living Shorelines: A Case Study of Habitat Enhancement," *Aquatic Conservation: Marine and Freshwater Ecosystems* 34, no. 3 (2024): e4118, https://doi.org/10.1002/aqc.4118.
32. Restore America's Estuaries, *Living Shorelines: From Barriers to Opportunities,* 5.
33. Ernie Franke, "Simple Aquascaping of Your Retention Pond Using Native Plants," *Bay Soundings: Tampa Bay's Environmental News,* December 21, 2012, http://baysoundings.com/simple-aquascaping-of-your-retention-pond-using-native-plants/.
34. Walter I. Priest III, "Practical Living Shorelines Tailored to Fit in Chesapeake Bay," in *Living Shorelines: The Science and Management of Nature-Based Coastal Protection,* edited by Donna Marie Bilkovic, Molly M. Mitchell, Megan K. La Peyre, and Jason D. Toft (CRC Press, 2017), 186.
35. C. Rodríguez-Calderón, "Forecasting Ecosystem Service Capacity: Effects from Sea Level Rise and Management Practices, Chesapeake Bay, Virginia," PhD diss., College of William and Mary, 2014.
36. Bas W. Borsje, Bregje K. van Wesenbeeck, Frank Dekker, et al., "How Ecological Engineering Can Serve in Coastal Protection," *Ecological Engineering* 37, no. 2 (2011): 113–22; Kevin L. Erwin, "Wetlands and Global Climate Change: The Role of Wetland Restoration in a Changing World," *Wetlands Ecology and Management* 17 (2009): 71–84.
37. Barry et al., "Perceived Effectiveness."
38. Rachel K. Gittman, Alyssa M. Popowich, John F. Bruno, and Charles H. Peterson, "Marshes With and Without Sill Protect Estuarine Shorelines from Erosion Better Than Bulkheads During a Category 1 Hurricane," *Ocean and Coastal Management* 102, pt. A (2014): 94–102; Barry et al., "Perceived Effectiveness."
39. Barry et al., "Perceived Effectiveness."
40. Gibson Marine Construction, "How to Know When Your Seawall Needs Repair or Replacement," January 25, 2012, https://gibson-marine.com/2012/01/25/how-to-know-when-your-seawall-needs-repair-or-replacement/; Seawalls Unlimited, "What Seawall Maintenance Might I Need to Do?," accessed February 1, 2025, https://www.seawallsunlimited.com/kind-seawall-maintenance-might-need/; Everlast Synthetic Products, "FAQ:

Seawalls Along the Florida Coastline," accessed February 1, 2025, https://everlastseawalls.com/florida-seawalls-faq/.

41. Gittman et al., "Marshes With and Without Sill."
42. Rowan Jacobsen, "Rebuilt Wetlands Can Protect Shorelines Better Than Walls," *Scientific American,* April 1, 2019, https://www.scientificamerican.com/article/rebuilt-wetlands-can-protect-shorelines-better-than-walls/.
43. Sarah Noel, "How Much Does a Seawall Cost?," HomeGuide.com, April 5, 2023, https://homeguide.com/costs/seawall-cost.
44. Timothy K. Blankenship, "To Repair or Replace a Bulkhead: That Is the Question," *Marina Dock Age,* April 2004, 22–24, https://www.coastalsystemsint.com/pdf/Media/Bulkhead_Converted.pdf.

# 4

# Living Shoreline Options for Florida

**THIS CHAPTER COVERS THE** various types of living shorelines that are available to Florida homeowners and other shoreline managers as well as the factors to be considered when designing a living shoreline for waterfront properties.

## Factors to Consider When Choosing a Living Shoreline

Determining what type of living shoreline is needed depends on the following:

- the degree of exposure to and erosion from waves and/or currents,
- the physical characteristics of the site (slope, depths, shoreline width, wind and wave exposures, tidal ranges, sediment types, existing man-made structures), and
- the location's environmental and biological characteristics (salinity, plus existing flora and fauna).

### The Degree of Erosion of the Shoreline

How exposed is the site and how severe is the shoreline erosion? Are there existing structures like buildings, swimming pools, fences or walls, or landscape features such as trees and gardens that are at risk if erosion continues? Is erosion being caused by storms, persistent or seasonal winds, or boat wakes? What is the erosion "history" of the shoreline? Has there been significant shoreline loss over time with increasing threats to the property and the assets on it from coastal flooding or waves? Have long-term residents noticed changes in the waterfront over the years? Older aerial photos compared with current ones may help to answer these questions. Even if erosion has not been notable in the past, an intervention with a living shoreline may be desirable to reduce future damage to the property.

## Slope and Bank Characteristics

Does the shoreline have a gradual slope of 1:10 or gentler (1 foot or less of vertical change every 10 linear feet), or a steep slope with a 3:1 (3-foot vertical change every 1 linear foot)? Is there scarping (a vertical or near vertical drop-off in bank height caused by rapid erosion)? A more gradual slope along a more protected shoreline may be immediately suitable for plantings to become established, as wave energy will be more dissipated by the time it reaches the plants. A more gradual slope indicates a wider intertidal zone with more opportunities to install living shoreline features in the proper locations; and these features will then also have more room to expand once the living shoreline flora and fauna become established. Steeper slopes or shorelines with moderate to heavy escarpments may need to be regraded to a gentler slope before planting. Terracing, or creating a series of platforms that gradually step down the bank, is another option to consider for steep slopes.

### SEDIMENTS

Sediment can consist of larger grain sizes like sand, or it can be finer like silt, mud, or clay. Sediment type and size may be a result of the typical wave energy for a particular site. In general, larger sediment particles are associated with greater energy and vice versa. (See table 4.1 for sediment types and sizes.) This is because, as shorelines are exposed to more wave action and currents, finer sediments are resuspended and exported, while on lower energy shorelines, these finer particles remain undisturbed. While the energy regime will usually be the strongest determinant of sediment grain size, other activities or events can also influence the type of sediments found on a shoreline. Recent dredging activities nearby might have led to finer sediments settling on a shoreline. Extreme changes in water quality, such as sudden, large-scale freshwater discharges, can leave behind finer sediments than what would otherwise be found at a particular location.

The predominant sediment type can affect the success of living shoreline projects in a few different ways. Sandy shorelines are well-drained and will dry out faster than areas with less sand content, potentially putting newer plantings

**TABLE 4.1** Sediment size classification according to the International Organization for Standardization (ISO)

| Name | Size range (mm) | Size range (approx. inches) |
|---|---|---|
| Coarse gravel | 20–63 | 0.8–2.5 |
| Medium gravel | 6.3–20 | 0.25–0.8 |
| Fine gravel | 2.0–6.3 | 0.08–0.25 |
| Coarse sand | 0.63–2.0 | 0.025–0.08 |
| Medium sand | 0.2–0.63 | 0.008–0.025 |
| Fine sand | 0.063–0.2 | 0.0025–0.008 |
| Coarse silt | 0.02–0.063 | 0.0008–0.0025 |
| Medium silt | 0.0063–0.02 | 0.00025–0.0008 |
| Fine silt | 0.002–0.0063 | 0.00008–0.00025 |
| Clay | ≤0.002 | ≤0.00008 |

SOURCE: ISO, "Geotechnical Investigation and Testing—Identification and Classification of Soil; Part 1: Identification and Description," ISO 14688-1 (2017), https://www.iso.org/standard/66345.html.

at risk of desiccation, especially in higher elevation zones. In this kind of situation, plan for the regular irrigation of new plantings.

Installed sill or reef structures could subside, or sink, in softer sediments. (See sidebar "Case Study, Project #2: A Natural Living Shoreline.") Finer sediments such as mud and silt are also often associated with higher nitrogen content, which may stimulate plant growth (see below) but may also negatively impact water quality. Finer sediments are more easily resuspended by waves and currents, blocking sunlight for seagrasses and other bottom-growing (benthic) vegetation in the area.

## The Location of the Shoreline Property

The location of a property will determine the most appropriate types of living shoreline structures and plants. For example, if a property is in an area that is prone to flooding, then one should plant tough salt marsh or riparian (riverine edge) plants that will be more likely to survive and provide the needed protection. The salinity ranges of the location in tidal waters (bay, estuary, or river) will also be an important factor to consider when making plant choices and when selecting hard materials for offshore structures. For the property owner, salinity can be easily measured with a simple hydrometer purchased from an aquarium store or a similar provider. There are also more precise and more expensive instruments to measure salinity. Ideally, salinity should be measured at different times of the year to obtain an estimated range of salinities that a site may experience during both wet and dry seasons. In general, properties near the mouth of an estuary (closer to the Atlantic Ocean or the Gulf of Mexico) will have a higher salinity range in the water than locations farther upstream, toward its head (river, freshwater source). The range of plant species that can be planted along the shoreline will change with the water's salinity. The suggested plant lists in appendix A indicate which plants are salt-tolerant. In general, the number of potential species will increase as the estuary location changes from saltier (more oceanic) water to fresher water. There are a more limited number of plant species (mangroves and cordgrasses being prime examples) that can tolerate daily inundation by saltwater.

Oysters also have a preferred salinity range, with their optimal salinity range generally between 10 and 28 ppt (parts per thousand). For reference, ocean salinity is between 34 and 36 ppt, while freshwater's salinity is 0 ppt. So, if an oyster reef is a desired component of a living shoreline, knowing the range of salinities for that property is important. Beyond the direct measurement of salinity, attention to presence of existing species and habitats either directly on the shoreline property or on adjacent or nearby locations may give a good idea of what can be successful.

Another factor when assessing the location of the shoreline property is the size of the buffer area. The buffer is the distance between the edge of the shoreline (generally demarcated by the average or mean high water [MHW] line) and the house, other buildings, or infrastructure on the property. A wider buffer area increases protection and offers opportunities to enhance the resilience, ecological function, and natural beauty of the area above the shoreline edge. (See chapter 8 on upland buffers.) An upland buffer and a living shoreline can work in concert to give the overall property increased resilience to coastal flooding events.

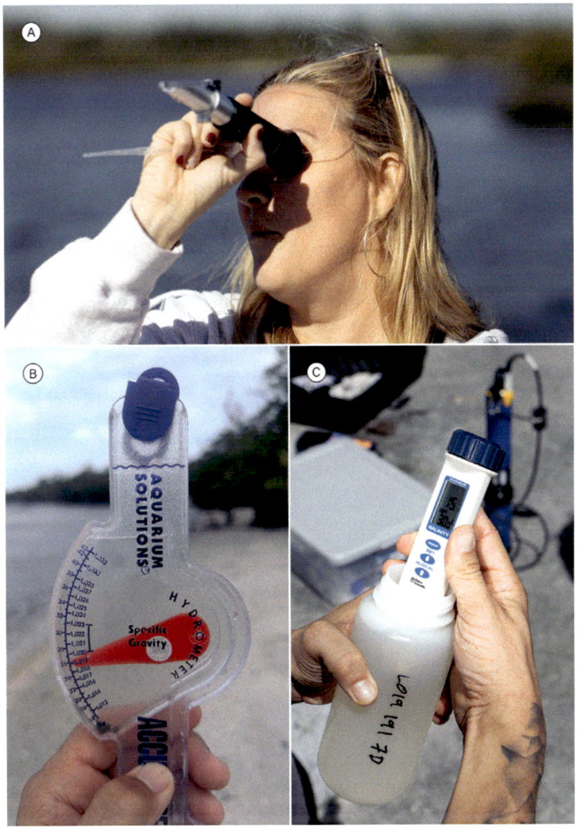


**FIGURE 4.1** Three different tools for measuring salinity: *(A)* refractometer, *(B)* hydrometer, and *(C)* electronic water quality meter. *Photos A and C, UF/IFAS photos by Tyler Jones. Photo B, UF/IFAS file photo by Vincent Encomio.*

## Setting Goals

The main objectives when considering a living shoreline may vary, but they most often include reducing erosion, improving water quality, and providing habitat for wildlife. Determining the main goals for a project will help to narrow down the options for the most appropriate living shoreline for that property.

Goals for restoration and stabilization may include some or all of the following:

To increase resiliency:

- Protecting upland property from existing erosion and future erosion
- Reducing damage from flooding due to king tides and rising sea levels

- Replacing or bolstering a failing seawall
- Reducing wave energy reaching the shore
- Accomplishing property protection goals at a lower cost compared to traditional methods

To improve the environment:

- Retaining or improving the habitat values of the shoreline—both above and below the MHW line
- Improving local water quality

To increase enjoyment and use:

- Improving access to the shoreline for recreational activities such as boating, fishing, watching birds or other nature, and paddle sports (kayaking, canoeing, paddle boarding)
- Enhancing the natural look of the shoreline, or "seascape"
- Masking an undesirable view
- Creating privacy

Obviously, homeowners want to protect their properties for safety and to preserve their most important economic investment—the home. Communities and other larger shoreline managers may want to improve access to public shorelines for the local population. Other goals are more rooted in personal values, desires, and tastes, especially regarding the environment. Many Floridians enjoy outdoor and water-based recreation such as fishing, boating or paddling, or birdwatching, and a healthy seashore environment is essential to participate in any of those activities.

Beyond protecting nature for their own personal enjoyment, most Floridians also have larger concerns for the health of the environment and the health of its animals and plants, and these are, in turn, linked to property values and economic health. Coastlines are changing in Florida, and unfortunately, many areas have lost shoreline habitats over time due to unsustainable land practices, excessive armoring, sea level rise, and stronger and more frequent hurricanes.

Some shoreline interventions can halt or reverse the effects of these impacts on specific shoreline properties, and topics in this book will provide insight into and inspiration on how to achieve protection goals for waterfront properties in a cost-effective manner while retaining or even enhancing the natural environment and its valuable functions.

## Budgeting

Living shoreline installations can range in cost from a few hundred dollars to several thousand dollars, but they generally cost much less than the armored alternatives. Homeowners in Florida reported in a recent research survey that living shorelines were 8.5 times less expensive to install and 2.7 times less expensive to maintain compared to armored shorelines.[1] It is important to factor in the cost of permitting, materials, installation, and maintenance when choosing a living shoreline. It is a good idea to get multiple quotes for a project from experienced contractors.

Costs will increase with the severity of erosion on a property, because greater intervention such as breakwater-style reefs made of rock, concrete, or oyster shells may be required to halt it. The homeowner may choose to install a living shoreline in stages to spread the expenses over several years. Site conditions may dictate how effective that approach will be. If erosion is severe and if the causes are ongoing, then a more costly breakwater reef may be needed first to protect less expensive, but more vulnerable components such as plantings in the intertidal zone. Installing appropriate plants above the shoreline or MHW line may be a less expensive approach and require fewer permits. This could be the first phase of either a living shoreline or a buffered shoreline. (See chapter 8 for more on buffered shorelines.) Some species such as red mangroves, certain marsh grasses, and vines tend to grow toward the water from just above the MHW line, especially if the shoreline is growing out or widening.

Even if there is not severe erosion on the shoreline, it is wise to act before conditions deteriorate. Sea levels are rising and will continue to impact Florida waterways for decades to come. A proactive approach that incorporates nature in the shoreline stabilization plans will contribute to its long-term resiliency while improving the health of the local environment. A resilient, self-sustaining shoreline will also lead to increased cost savings if more expensive "gray" infrastructure (rock or concrete-based infrastructure) can be avoided.

## Types of Living Shorelines

There are different types of living shorelines. Options run from a true living shoreline consisting only of natural, or "green," elements (native plants and natural materials to construct sills or reefs) to hybrid living shorelines that combine both

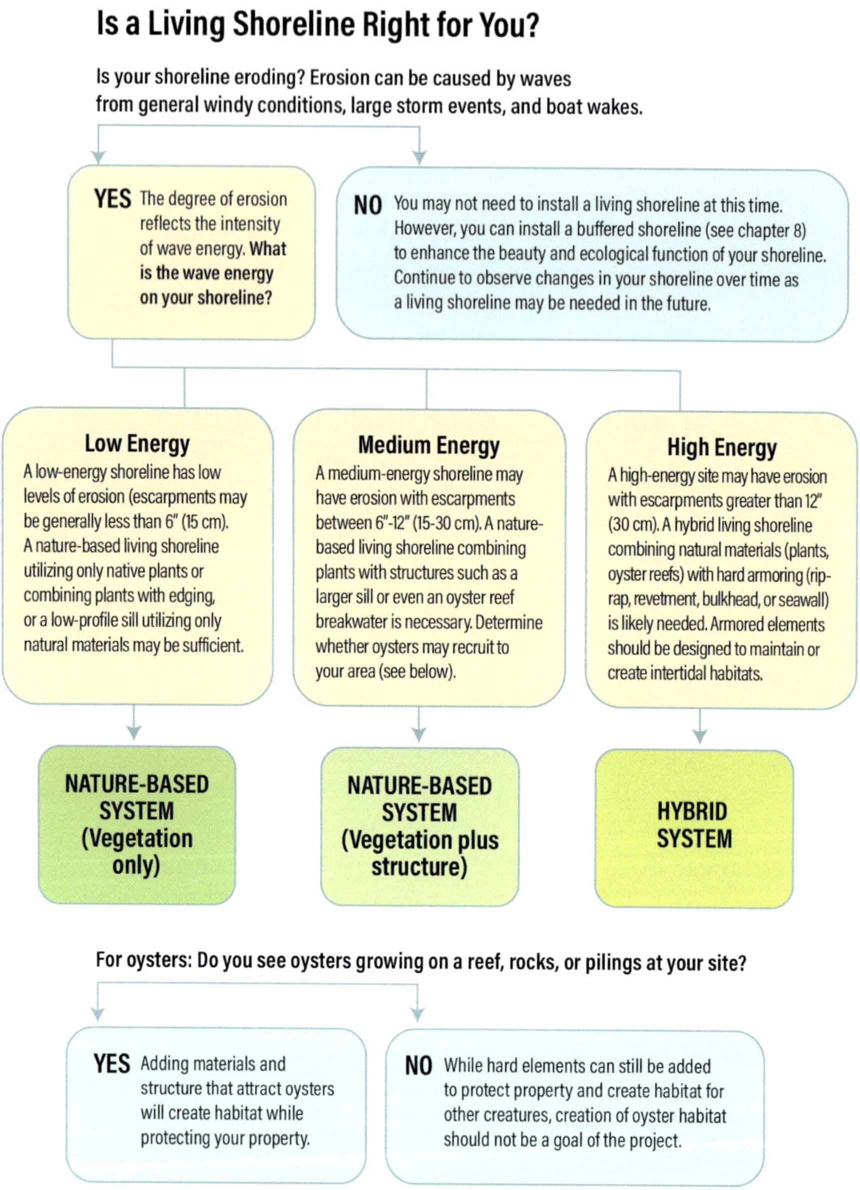


FIGURE 4.2 A decision tree can assist the project team in determining what type of living shoreline and what activities should be planned as part of the project (e.g., planting, regrading, reef installations, etc.). Following the guidelines as laid out in expedited permit types (e.g., state exemption or Nationwide 54) can save time and money but might not be appropriate for all projects. *Illustration by Marjorie Shropshire.*

natural, or green, infrastructure with hard (rock or concrete-based), gray infrastructure. The degree of each (green or gray infrastructure) will depend on the site conditions and the goals of the property owner. The existing site conditions are especially important as different regions and locations in Florida may have varying proportions of existing green and gray infrastructure, or combinations of both. For example, in the Panhandle, there may be large landscapes of salt marsh that would integrate well with a full living shoreline approach.

Conversely, in South Florida, many areas already contain a high proportion of gray infrastructure and automatically necessitate a hybrid approach to achieve a living shoreline by incorporating green elements into this existing infrastructure. The degree of gray infrastructure in a hybrid living shoreline may also depend on the proximity of a house or other significant human infrastructure, such as swimming pools, to the shoreline. Oftentimes, especially in Florida, homes are built as close to the shoreline as possible. Seawalls are often built at the MHW mark. When homes are built close to seawalls, bulkheads, or riprap, the available intertidal shoreline space is vastly truncated or even eliminated altogether in some sea-walled properties, making these properties more vulnerable to storms and sea level rise.

Regaining a gradual, more natural transition between intertidal and upland zones is one goal of living shoreline projects. The widening of shorelines will provide both protective and ecological benefits. Hybrid living shorelines may be necessary to reconcile the impacts of development with nature.

This chapter covers how to determine what kind of living shoreline is appropriate for a specific site and provides options for two different categories of living shorelines that can be installed.

## Natural Living Shorelines

A natural living shoreline uses only natural, nature-based elements (plants, natural rock, wood, fiber materials, oyster or other bivalve shells) as building materials. A natural living shoreline will also mimic and integrate, in appearance and function, the natural coastal environment such as a salt marsh, mangrove forest, and oyster reefs. It may also be directly connected or adjacent to the surrounding natural environment, as in the case of rural shorelines adjacent to natural, relatively undisturbed coastal habitats. But more urbanized settings near to or adjacent to coastal preserves also may also afford opportunities to achieve a natural living shoreline.

Natural living shorelines are more likely to be built in low-energy settings, where waves are small and the land slopes gently to the water, or in protected locations. They mimic natural systems with plants generally associated with coastal wetlands (i.e., mangroves and cordgrasses). Many low-energy shorelines may

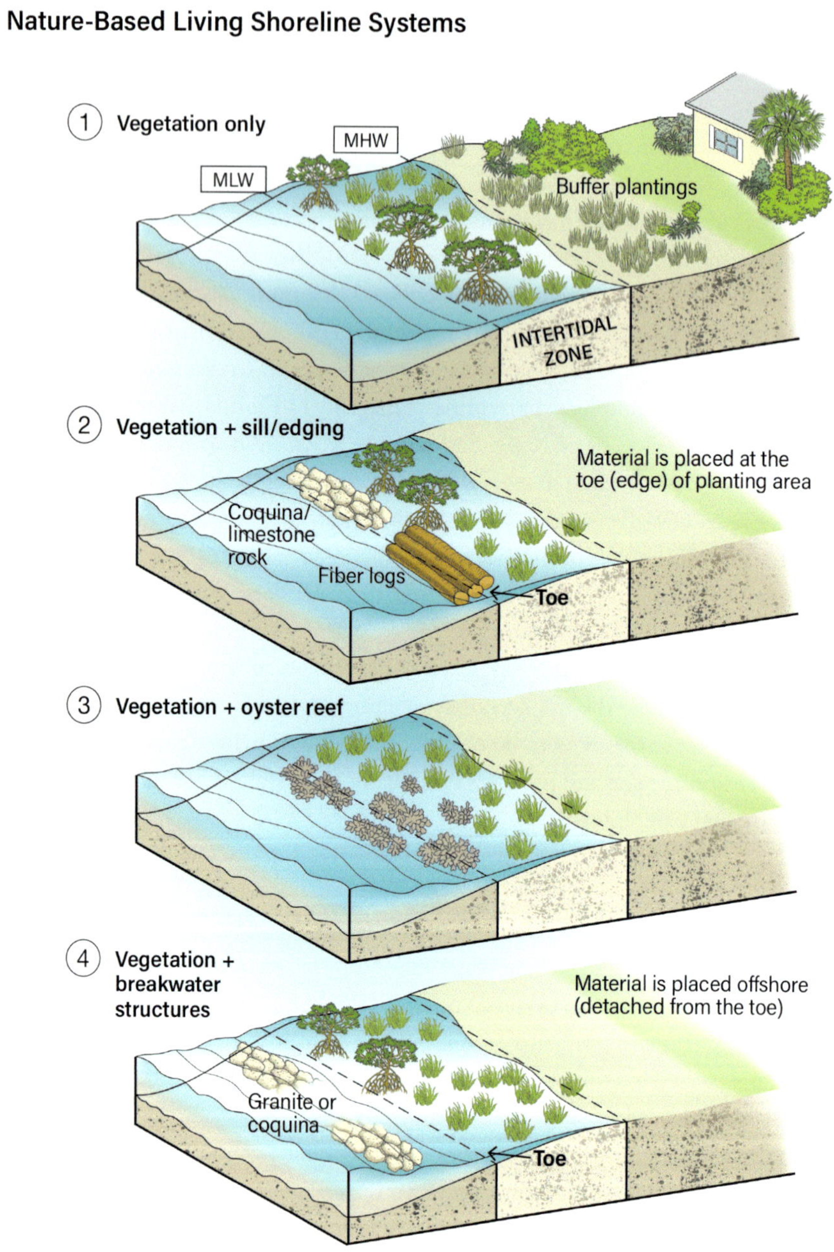


**FIGURE 4.3** Nature-based living shoreline systems use only natural materials. Options range from vegetation only to vegetation combined with an oyster reef or edging and sill materials. *Illustration by Marjorie Shropshire.*

require only the addition of vegetation to stabilize the shoreline. If there is enough wave energy to require some protection for plantings, oyster reefs can be constructed to mimic natural oyster habitat. The site conditions (presence of existing oysters, optimal salinity ranges, shallow depths, non-muddy sediments) must be appropriate, however. If the location will not support oyster habitat, and if protection is still needed, then structures made of natural materials (fiber logs, limestone, or coquina rock) can be installed to protect plantings. If there is little to no erosion on the shoreline, the property owner might not need to do anything aside from adding plants and perhaps an oyster reef to improve the habitat, clean the water, add interest, and increase the shoreline's beauty.

## Living Shoreline Sills and Breakwaters

Structures to protect plantings are generally categorized as edging, sills, or breakwaters. Both edgings and sills are defined as features connected with and part of the toe (the outer edge of planting area, which is near the mean low water [MLW] line) of the existing slope of the planting area. A breakwater is defined as an offshore structure, detached from the toe of the slope.

A sill is a small-scale structure (generally less than one foot in height) that is placed contiguous to the planted area. Sills provide immediate protection for living shoreline plantings located upslope from the sill. Sills can be constructed using natural fiber "logs" (made of jute, coconut fiber [coir], or other natural fiber and then filled with the same fiber or compostable materials) that are staked down onto the surface. In addition, plants can be directly installed within the fiber log, which helps to keep them in place within the living shoreline. While the fiber logs could be removed once plants are established, removal is not necessary or recommended because they will eventually bio-degrade and help stabilize the planting area. A sill tends to trap sediment behind it (or toward the upland area), which further stabilizes the planting area and makes it more conducive to additional plant establishment.

Sills are more commonly constructed of harder material, such as shell or rock, as wave energy on most shorelines is somewhat higher and exceeds the capacity of a fiber log to stay in place, even if staked. In these cases, a sill structure could be used to create an oyster reef as well. In that case, the sill should be constructed of oyster shells, prefabricated concrete units, or natural rocks that encourage oyster settlement, especially if there is natural oyster recruitment at the project site (usually evidenced by the presence of oysters on other surfaces nearby). The edge of the sill must be placed in the proper intertidal location or water depth for oysters. If the sill is not in an area conducive to oyster propagation, then oyster shell is

## CASE STUDY

### PROJECT #2 OF A NATURAL LIVING SHORELINE

**Location:** Southeast Florida, Martin County, Jensen Beach, Indian River Lagoon

**Upland Property Ownership**: Private business (Driftwood Motel)

**Site History:** This shoreline experienced continuing erosion over time, most notably after the 2004 Hurricanes Frances and Jean. The owner was concerned about historical erosion as the historical mean high water mark was significantly waterward from the current MHW. Prior to 1990, concrete rubble was deposited on the shoreline to stabilize it. The site had an existing, small oyster reef (~100 feet$^2$ [9.3 m$^2$]) adjacent to the motel's dock. The dock pilings and nearby concrete rubble were colonized by oysters. The oysters in this area were numerous and healthy, indicating that oyster reef construction had high potential for success.

**Dates of Construction:** Spring 2016 through 2018.

**Living Shoreline Type:** Natural

**Living Shoreline Building Blocks:** Oyster reefs (bagged shell), smooth cordgrass (*Spartina alterniflora*), and seagrass (*Halodule wrightii*) planted.

**Size:** Shoreline length: 70 feet (21.3 m); 10 reefs, each 50 square feet (4.6 m$^2$) for a total of 500 square feet (46.5 m$^2$); planting area ~400 square feet (37.2 m$^2$)

**Permitting:** A Florida Department of Environmental Protection (FDEP) individual Environmental Resource Permit was awarded in 2015. A US Army Corps of Engineers Nationwide Permit 27 for Aquatic Habitat Restoration, Enhancement, and Establishment was issued in 2016. The permits allowed oyster reef and seagrass installations. Marsh grass plantings were installed above and at the MHW line and, thus, did not require a permit.

**Construction:** Oyster reefs were constructed first. Low marsh vegetation (*Spartina alterniflora*) was installed in ~2017 above and at the MHW line. Intertidal oyster reefs were constructed from bagged, recycled shells. Each reef was 1–1.5 feet high (0.5 m) (2 tiers of bagged shell). The reef designs were curvilinear in shape, each ~10–15 feet (3–3.6 m) long and 4–5 feet (1.2–1.5 m) wide.

**Maintenance:** Replanting was done when and where cordgrass washed out due to waves and king tides. Otherwise, there was very little maintenance.

**Cost:** Plants were obtained for free from the Marine Discovery Center's donor marsh. Costs included shell bags and were related to transportation and labor. Overall costs were ~$3,000–$5,000.

**Notes:** The motel owner worked with a local nonprofit (Florida Oceanographic Society [FOS]) to install this living shoreline and was motivated by her concerns for declining water quality and loss of habitats like oyster reefs and seagrass due to freshwater discharges from Lake Okeechobee. With this site and four other properties a few miles south, FOS's goal was to create estuarine habitats downstream from Lake Okeechobee discharges to improve the southern Indian River Lagoon's resiliency to discharge events. The cordgrass plantings and oyster reefs were successful and remain well-established and self-sustaining. Grasses and sand have covered most of the concrete debris. Cordgrass plantings have migrated and spread from their initial planting locations above MHW line to extend into low marsh areas and now abut the closest inshore oyster reefs.

**Challenges:** The most persistent challenge for this site has been high water levels and king tides, which typically occur in the fall months. Initial cordgrass plantings were washed out or drowned and had to be replaced. Although high water likely still causes plant stress, the cordgrass plants now largely extend above the highest water levels. High water levels likely impact the oyster reefs as well by limiting intertidal exposure for about 2 or 3 months per year. At the time of this writing, both the cordgrass plantings and oyster reefs are well-established and resilient. The living shoreline remained intact and resilient in storm conditions generated by Hurricanes Matthew (2016), Irma (2017), and Ian and Nicole (both 2022). Hurricane Irma did, however, bury initial seagrass plantings. Up to that event, planted seagrass was healthy and growing.

FIGURE 4.4 A living shoreline at the Driftwood Motel, Jensen Beach, Florida. *(A)* Shows placement of small oyster reef breakwaters prior to planting. *(B)* Shows the extent of smooth cordgrass (*Spartina alterniflora*) growth three years later. The plants have migrated waterward toward the first line of oyster reefs indicating successful expansion of the shoreline especially when compared to the adjacent shoreline with only riprap. *Photos by Owen Silvera, Georgia Institute of Technology.*

probably not an appropriate material to use for a sill. (See chapter 5 for more details.)

Small to medium-scale breakwaters can be constructed from approved materials such as granite, limestone, recycled clean concrete, or coquina. As with sills, a breakwater can also be constructed to create oyster habitat, using shells, rock, concrete units, or some combination of these. Breakwaters for living shorelines are distinguished from traditional gray infrastructure breakwaters in that they are smaller and allow the flow of water around and behind them. They are built for the dual purposes of protecting plants by reducing wave energy and providing habitat for marine life. Natural materials specifically for habitat provision also distinguish living shoreline breakwaters from traditional ones. (See chapter 6 on permitting.)

## Hybrid Living Shorelines

Hybrid approaches may be necessary when the wave energy at a particular shoreline is too high for a fully natural living shoreline. Higher-energy shorelines are caused by increased wave exposure due to increased fetch and wind, deeper water closer to shore, frequent wakes from boat traffic, or a combination of these factors. A hybrid approach allows the incorporation of living shoreline building blocks to enhance ecological function and to augment existing infrastructure with natural, self-sustaining habitats.

Hybrid systems include man-made structures like seawalls and bulkheads or larger breakwaters in addition to plants and other natural materials. While adding armoring to a shoreline is not recommended if a nature-based living shoreline can be constructed instead, some shorelines may need to be stabilized with a hybrid system. For example, medium to high-energy shorelines, those with a less gradual slope (and limited space to extend that slope to a more natural degree), and areas subject to frequent boat wakes may not be suitable for full nature-based systems; but they can be stabilized with a hybrid shoreline design. Also, where there are existing seawalls, hybrid systems can be created by adding green elements to retrofit them, which provides extra protection and increased environmental function.

There are also many man-made waterways, such as canals, that share many of the same challenges as other shorelines (structures built out to the edge of seawalls, vulnerability to sea level rise and coastal flooding, plus the disruption of the intertidal zone and its ecological processes). The same solutions of hybrid living shoreline approaches apply to canals as well. Homes on canals that have seawalls can be retrofitted outside the seawall with riprap (rocks) or planter boxes, and they can be planted with mangroves and marsh grasses. Retrofitting can help stabilize failing

## Hybrid Living Shoreline Systems

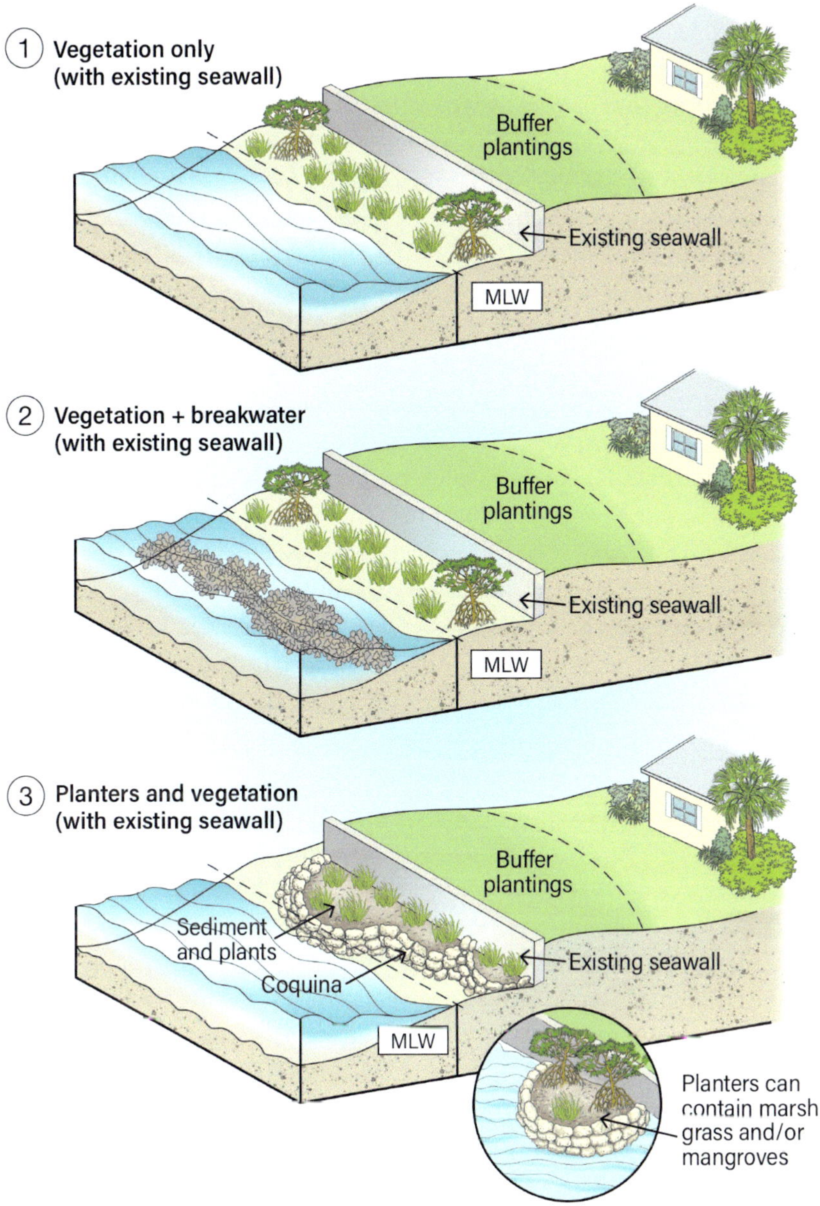


**FIGURE 4.5** Hybrid living shoreline systems combine hard infrastructure with living shoreline elements and can be used to retrofit existing seawalls. Options range from cultivating vegetation adjacent to the seawall to installing vegetation plus breakwaters or planters. Usually, the hard infrastructure is already in place when the living shoreline elements are added. These projects are referred to as living shoreline retrofits. *Illustration by Marjorie Shropshire.*

seawalls and is a less expensive solution than a full seawall replacement. These "redeemed" hardened structures can be incorporated into a hybrid living shoreline to reduce erosion and regain intertidal habitat without a drastic intervention that may include removal of existing wall or rock structures and possible fill.

### AUGMENTING SEAWALLS AND BULKHEADS WITH LIVING SHORELINES

If there are shallow areas in front of or on the water side of a seawall or bulkhead, vegetation can be planted there. As sediment is trapped by the vegetation, the plants (marsh grasses or mangroves) may extend waterward, which will ultimately expand the width of the shoreline and its intertidal zone. In higher-energy environments, a sill or breakwater structure may be needed to protect the new plantings and allow them to establish themselves. A successful living shoreline in this example can protect not only upland areas but also the seawall itself, ideally rendering it obsolete over time as shoreline builds up in front of it. The existing depths and tidal ranges must be able to support plantings in front of the seawall, however. Sites with exposed areas at low tide would be the best candidates for this approach, as low marsh vegetation tolerates regular inundation but cannot survive in areas that are never exposed to air through the fluctuation of the tides. In many shorelines with seawalls, the depths are too great to allow for this. Also, seawalls can sometimes cause erosion in front of the structure, and in these cases, mangrove planters may be the appropriate approach.

### GREENER SEAWALLS

Seawalls can be modified to create more natural areas. These enhanced seawalls (aka living shoreline retrofits) can be incorporated into a larger living shoreline design that includes introducing a more gradual slope using riprap containing vegetation that is either directly planted (e.g., mangrove planters) or encouraged by the creation of a more favorable environment for the natural recruitment of appropriate native plants. Outcroppings and indentations can be incorporated into the seawall construction, increasing habitat complexity and refuge space.[2] A similar approach was incorporated into a larger living shoreline retrofit project in Seattle, Washington.[3] These enhancements can be designed to be aesthetically pleasing. Mangrove Reef Walls (https://www.reefwall.com/) can also be installed over existing seawalls. These habitat panel inserts, manufactured to resemble red mangrove prop roots, can enhance three-dimensional habitat space for encrusting marine organisms like oysters, mussels, barnacles, and sponges. In 2022, concrete panels were used to create living seawalls in Sydney, Australia, to restore biodiversity to Sydney Harbor.[4] While these approaches are not technically living shorelines on their own, they can be an integral part of broader efforts to reestablish shoreline

**FIGURE 4.6** *(A)* In shallow areas, seawalls or bulkheads can be augmented with living shoreline vegetation, and sills or breakwaters can be added in higher energy zones to protect the plants. As the plants gather sediment and grow outward, the width of the shoreline and the intertidal zone increase. *Photo by Zachary Schang, Florida Department of Environmental Protection. (B)* A scawall retrofit with living shoreline elements on both the upland and waterward sides. The upland buffer of native grasses traps pollutants and slows water runoff, while the plants on the waterward side add habitat while also protecting the vulnerable toe of the seawall. *Photo by Annie M. Roddenberry.*

function, particularly in urban coastal areas where intertidal space is limited and shorelines are already armored.

## LIVING SHORELINES TO REDEEM FAILED SEAWALLS

Seawalls or bulkheads will eventually fail as their structural lifespans are finite. (See chapter 3 for more information.) Seawall lifespans are sometimes shortened by increasing storms and sea level rise. Enhanced erosion at the base of seawalls due to wave reflection is a characteristic of seawalls and leads to their eventual failure. Coastal flooding and storm surges over the top of a seawall cause pressure on the structure from behind it, which also destabilizes soils and leads to erosion on the landward side of that structure. These impacts can lead to failure in aging seawalls that are not repaired or maintained. Replacement of a traditional seawall

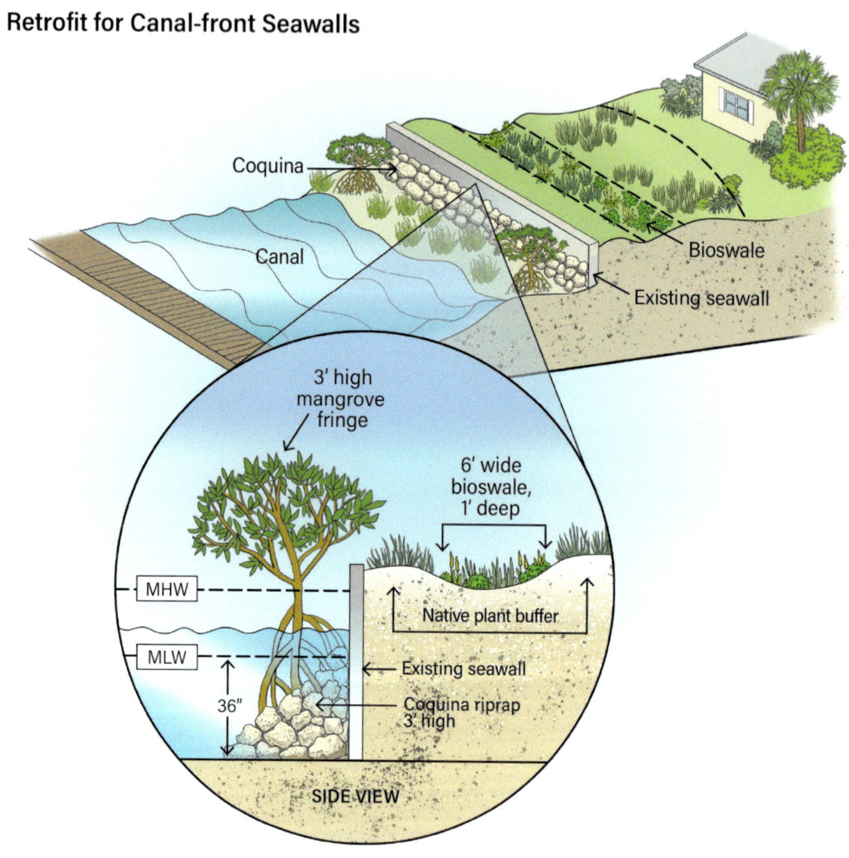


**FIGURE 4.7** A canal-front property with an existing seawall can be retrofitted with coquina riprap. In this example, mangroves are planted directly into the riprap, but another technique is to add mangrove planters to the seawall. In addition to the elements on the water side of the seawall, a shoreline buffer of native plants, a bioswale, and a vegetated upland buffer provide additional protection to the seawall while helping to protect water quality. *Illustration by Marjorie Shropshire. Side view detail adapted from Nicholas Frank Sanzone, LagoonLife.org.*

can be an expensive option ($1,500–$3,500 per linear foot [0.3 m]).[5] Using natural infrastructure to "redeem" a failed seawall can be one way to create a hybrid living shoreline.

A reef breakwater or riprap can be installed waterward of the failing seawall or other marine infrastructure to help protect it. Plants can be installed behind these structures and even within the area of the failed wall if sufficient soil exists in that area.

The seawall could be removed as well to make way for a full natural living shoreline.[6] Understandably, there may be a lot of apprehension about doing this. Also, careful analysis of the wave energy at a site must be performed before considering seawall removal. It is not advisable along high-energy shorelines. There must also be a sufficiently sized upland buffer to allow for natural movement of the shoreline that does not threaten homes or other significant man-made structures. (See chapter 8.)

### MANGROVE PLANTERS

There may be room to extend riprap out from an existing seawall to allow for shoreline vegetation such as mangroves to recruit naturally or to be directly planted. Direct plantings within riprap may be challenging, however, and will likely require

**FIGURE 4.8** This large-scale mangrove planter, part of the Snook Islands Project in Lake Worth Lagoon, uses boulders to contain sediment where red mangroves were planted. An opening on one side allows water and wildlife to access the interior. *Photo by Ginny Stibolt.*

**FIGURE 4.9** Small-scale mangrove planters built adjoining a seawall are suitable for residential sites or areas with limited space available, such as this one at McNabb Park in the City of Cocoa Beach, Florida. *UF/IFAS photo by Holly Abeels.*

the installation of structures that can contain soil or sediment where the plants can grow. Another approach that has been employed is constructing riprap planting areas in which a semicircular containment area is built using appropriate materials such as rocks and boulders. The open area is then filled with sediment to create an area in which mangroves or other intertidal vegetation (smooth cordgrass and other salt marsh plants) can be planted. This type of hybrid living shoreline has been used in South Florida for large-scale projects on shorelines longer than 100 feet (30.5 m), but it also has potential at smaller scales such as residential properties.

**TABLE 4.2** Types of shoreline management methods that work well in different energy (wind, wave, fetch) climates

| Infrastructure type | Living shoreline (LS) methods | Energy |
|---|---|---|
| Green | Vegetation only | Low |
| Green | Vegetation + edging | Low |
| Green | Vegetation + sill | Low to moderate |
| Green | Vegetation + breakwater/oyster reef | Low to high (height*, width, and material strength will increase with shoreline energy) *Height may be limited by permits to be considered a LS structure |
| Hybrid | Retaining wall + plants | Low to moderate |
| Hybrid | Terracing + plants and/or oysters (sills or reefs) | Moderate |
| Hybrid LS | Seawall + mangrove "planter" (variation of previous) | High |
| Hybrid/Gray | Seawall + habitat inserts | High |
| Hybrid/Gray | Bulkhead/seawall + riprap | High |
| Gray | Revetment/riprap | High |
| Gray | Bulkhead/seawall | High* (*bulkheads and seawalls can be in low-energy canals) |

*Note:* LS = living shoreline.

Design should consider the appropriate tidal height of the riprap structure and the planting area for shoreline vegetation. Effective planting of red mangroves is recommended around or just above the MHW line at a site. Also, the size, particularly the waterward extent of the planter structure, may be limited by the seawall's distance to navigable waters or structures such as docks. These constraints are especially apparent in canals. A riprap planter may be a viable, and sometimes the only, option to achieve a hybrid living shoreline that reconciles nature with the built environment. This type of hybrid approach can increase the resiliency of highly urbanized Florida cityscape waterfronts to coastal flooding due to storms, king tides, and sea level rise.[7]

If possible or necessary, most of these infrastructure types—except for the traditional bulkhead or seawall—can be combined with a detached breakwater/oyster reef.

## Conclusion

Living shorelines are an effective, sustainable, and long-lasting method for protecting coastal properties from erosion and other coastal hazards. They also provide other benefits, but they must be designed appropriately for the site conditions. Before installing a living shoreline on a Florida waterfront property, consider the site factors and the various options for living shorelines to find one that is the best fit for the situation.

## Notes

1. Savanna C. Barry, Laura K. Reynolds, Anna E. Braswell, Rachel K. Gittman, Steven B. Scyphers, and Ashley R. Smyth, "Perceived Effectiveness Drives Shoreline Decision-Making for Florida's Waterfront Property Owners," *Ocean and Coastal Management* 258 (2024): 107353, https://doi.org/10.1016/j.ocecoaman.2024.107353.
2. Fabio Bulleri, and Maura G. Chapman, "The Introduction of Coastal Infrastructure as a Driver of Change in Marine Environments," *Journal of Applied Ecology* 47 (2010): 26–35.
3. Donna Marie Bilkovic, Molly M. Mitchell, Megan K. La Peyre, and Jason David Toft, eds. *Living Shorelines: The Science and Management of Nature-Based Coastal Protection.* (CRC Press, 2017), ch. 21.
4. Melanie J. Bishop, Maria L. Vozzo, Mariana Mayer-Pinto, and Katherine A. Dafforn, "Complexity-Biodiversity Relationships on Marine Urban Structures: Reintroducing Habitat Heterogeneity Through Eco-Engineering," *Philosophical Transactions of the Royal*

*Society B: Biological Sciences* 377 (2022): 20210393. https://doi.org/10.1098/rstb.2021.0393.

5. Florida Fish and Wildlife Conservation Commission, "2023 Living Shorelines Training for Marine Contractors," course manual. Courses have been coordinated by Florida Sea Grant (https://www.flseagrant.org/) since 2021.
6. Thomas Ries, "Living Shorelines and Seawalls Provide Nature-Based Solutions to Stabilize Florida's Coasts," *The Invading Sea: Florida and the Climate Crisis,* January 11, 2024, https://www.theinvadingsea.com/2024/01/11/living-shorelines-seawalls-sea-level-rise-flooding-nature-based-solutions-florida/
7. Alex Harris, "Mangroves vs. Seawalls? Mix May Be 'Best of Both Worlds' to Take on South Florida Sea Rise," *Miami Herald,* April 17, 2022, https://www.wusf.org/environment/2022-04-17/mangroves-vs-seawalls-mix-may-be-best-of-both-worlds-to-take-on-south-florida-sea-rise; Laura Parker, "Can Restoring Mangroves Protect Miami from Rising Seas? Living Shorelines Featuring Mangroves Can Stabilize Coasts, Limit Flooding, and Save Landowners Money," *National Geographic,* July 26, 2022, https://www.nationalgeographic.com/environment/article/can-restoring-mangroves-protect-miami-from-rising-seas.

## Resources

Barry, Savanna, Sara Martin, and Eric Sparks, "A Homeowner's Guide to the Living Shoreline Permit Exemption, Part 1: Florida Department of Environmental Protection," EDIS Publication SG187, UF/IFAS, July 24, 2023 (original publication February 2019), https://edis.ifas.ufl.edu/publication/SG187.

Florida Living Shorelines, https://floridalivingshorelines.com/.

NOAA Fisheries, "Understanding Living Shorelines," accessed February 3, 2025. https://www.fisheries.noaa.gov/insight/understanding-living-shorelines.

# 5

# Living Shoreline Building Blocks

**THIS CHAPTER COVERS NATURAL** elements and material options that can be used to build a living shoreline, including oyster reef substrates, oyster reef alternatives, bank-stabilization options, and plant species. Sections are organized from the water to the shoreline. See chapter 8 for information about building blocks for the upland zone.

It is best to use local materials to build living shorelines. For example, living shorelines in the southerly portions of Florida should include mangroves, while herbaceous saltmarsh plants work well for living shorelines in the northern portions (e.g., the Florida Panhandle). Also, guidelines on allowable or recommended materials may vary by state or region or change over time, so be sure to consult an environmental professional and/or regulatory agency (see chapter 6) before choosing the materials for a project.

*Note:* The inclusion of specific material in the descriptions below does not constitute an endorsement by the authors, and the authors have no financial stake in any of the materials covered. The authors aimed to include all available options and to provide a representative cross section of the types of materials available at the time of writing, but material availability and options will change. New materials come on the market often, and some options may have limited regional availability.

## Oyster Reefs: An Overview

Oysters are considered ecosystem engineers that can change or create habitats. In Florida, the eastern oyster (*Crassostrea virginica*) forms vertical reefs that are effective at absorbing wave energy in low-energy to medium-energy settings. A study using wave tank research showed that when an oyster shell substrate with live oyster recruitment was combined with one-year-old plantings of smooth cordgrass, wave energy was reduced by 67 percent.[1] Furthermore, oysters can keep pace with sea level rise, making them an advantageous partner for marshes when the goal is self-sustaining erosion control.[2]

Oysters are filter feeders that ingest algae, bacteria, contaminants, and sediment. One adult oyster can filter up to 50 gallons of water a day,[3] and this in turn improves water clarity. Clear and clean water is important for the growth of benthic organisms (those that survive at the bottom of the water) such as seagrasses and more. Individual oysters stick together to form complicated reef structures where other organisms can live. Just one square meter of oyster reef can provide as much as 50 square meters of habitat.[4]

BENEFITS OF OYSTER REEFS

- Protect the shore from storms by distributing wave energy
- Prevent erosion
- Improve water quality and clarity
- Create habitat for fish and other animals

Oyster reefs serve as habitats that provide nurseries, food, and homes for hundreds of species of animals, from fish to invertebrates. (For more information on oysters in natural habitats, see chapter 1.)

### Building Blocks for Oyster Reefs and Breakwaters

To build a new oyster reef, loose oyster shells can be collected and placed in long-lasting bags or cages; there are also many options for prefabricated concrete units. Units are lined up, interlocked, and/or stacked on top of one another to create a reef structure. Gaps of at least 5 feet (1.5 m) should be left between reef sections so that water, animals, and people can move around the reef. Often the reefs are crescent shaped or meander with the shoreline shape, this way, they are both functional and aesthetically pleasing.

Oysters, especially at higher salinities, can benefit from being out of the water a few hours a day, and constructed reefs are placed in the intertidal zone where they will not interfere with boating, access to docks, paddling, or wading.

Before installing an oyster reef on a living shoreline, check the site for evidence of existing oysters. If no oysters are present, the site might not be a good location for an oyster reef. The presence or absence of oysters in the shoreline location will be an important factor to consider when selecting reef substrate.

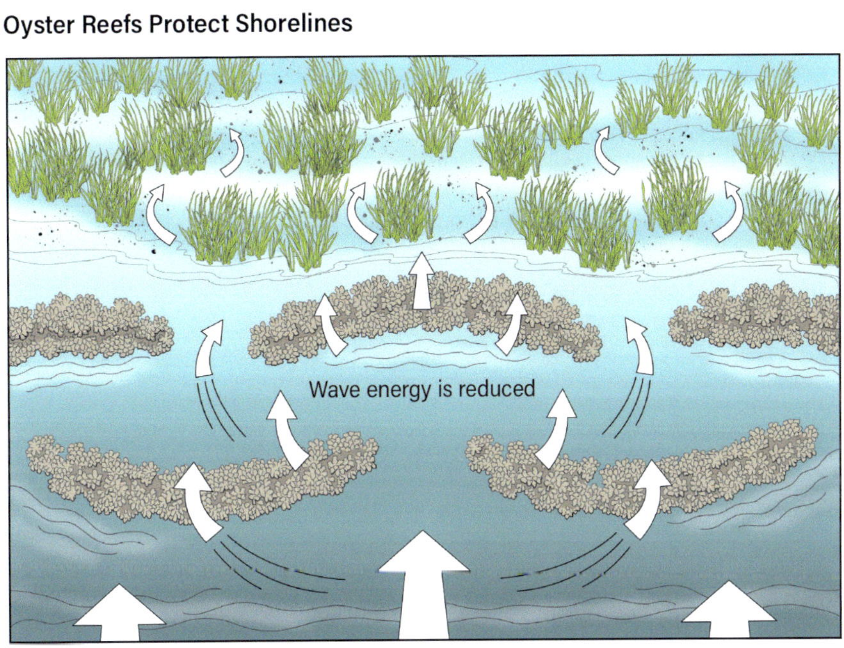


FIGURE 5.1 Oyster reef structures act as wave breaks, dissipating wave energy and protecting the shoreline. *Illustration by Marjorie Shropshire.*

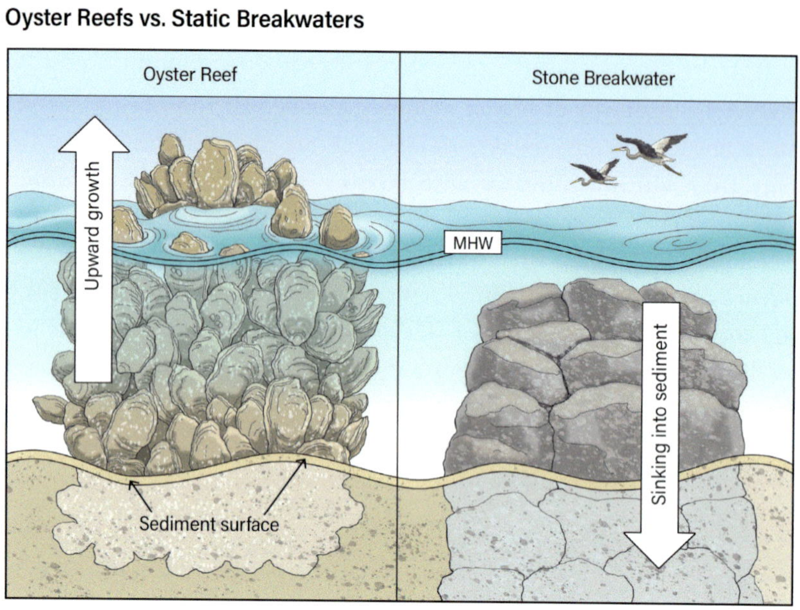


**FIGURE 5.2** Oyster reefs at work vs. stone breakwaters. Oyster reefs are living, growing, and dynamic. Their upward growth can keep pace with rising water. Static structures, like stone breakwaters, sink into the underlying sediment over time, so their effectiveness is lessened, and they cannot keep pace with rising water. *Illustration by Marjorie Shropshire.*

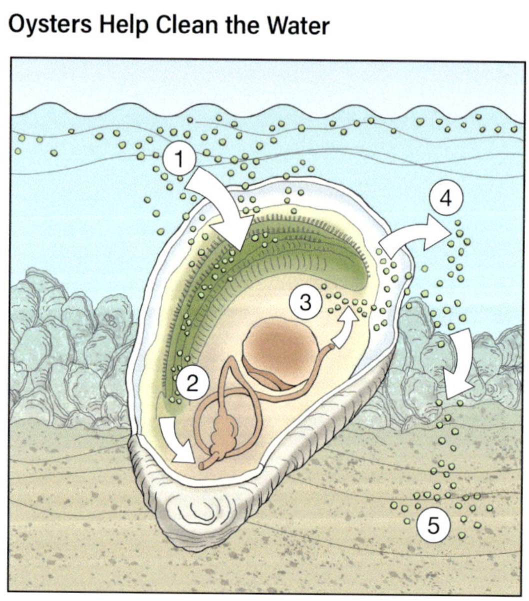


**FIGURE 5.3** Oysters help clean the water. *(1)* Oysters are filter feeders, taking in oxygen, plankton, nutrients, detritus, and fine sediments from the water. *(2)* Cilia on their gills move oxygen and food to the heart and digestive tract. *(3)* Wastes are ejected as feces. *(4)* Food waste is ejected as feces, but undigested materials are ejected as pseudofeces. *(5)* Both feces and pseudofeces sink to the bottom as biodeposits. *Illustration by Marjorie Shropshire.*

**Biodiversity of Oyster Reefs**

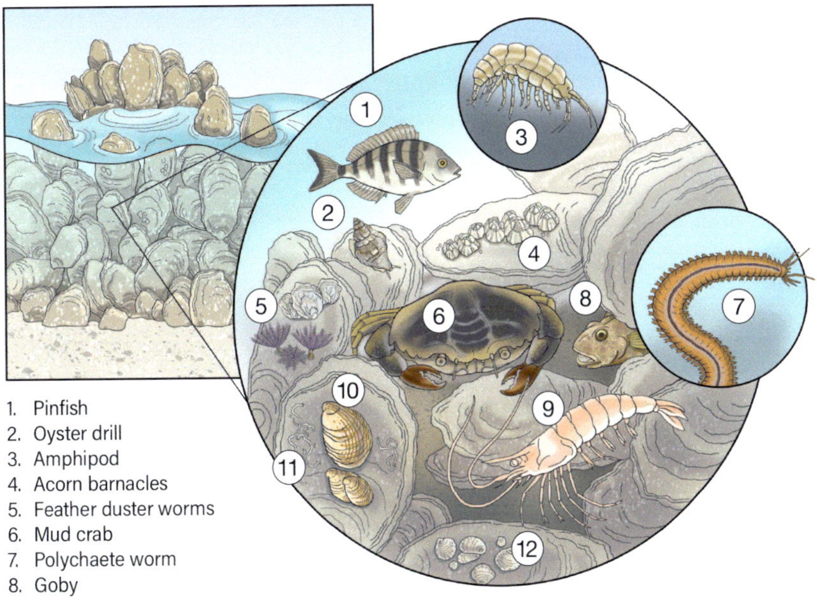

1. Pinfish
2. Oyster drill
3. Amphipod
4. Acorn barnacles
5. Feather duster worms
6. Mud crab
7. Polychaete worm
8. Goby
9. Shrimp
10. Slipper snail
11. Tube worm
12. Oyster spat

**FIGURE 5.4** Biodiversity of oyster reefs. The surface of an oyster reef is complex, providing numerous spaces for a diversity of creatures of different sizes and shapes to hide and feed. Larger underwater predators are attracted to the reefs to hunt. When oyster reefs are above the water for part of the day, the reef also attracts foraging birds and other animals. *Illustration by Marjorie Shropshire.*

## Reefs, Sills, and Breakwaters

A true reef is a ridge or shoal of stable material such as rock, coral, or oyster lying beneath the surface of a natural body of water. In contrast to sills and breakwaters, which are human-made, reefs result from natural processes such as layers of oyster recruitment over time, corals building their skeletons, or wave energy eroding rock outcrops. Reef restorations are an attempt to reinstate these natural processes.

A sill is built with materials such as stones or bio logs that are deployed along the waterward extent (toe) of a vegetated zone to reinforce and protect the area from moderate-energy waves and currents. Sills differ from breakwaters in that they are constructed right next to the vegetated area.

A breakwater is a structure built to provide a barrier to or protection from waves. There are many definitions of the term *breakwater,* but usually they are offshore structures designed to reduce incoming wave energy, resulting in calmer water between the shoreline and the breakwater structure. For high-energy sites, working with a coastal engineer is strongly recommended, as they can perform a Coastal Conditions Analysis that will inform the site design's breakwater orientation, height, and material density to withstand the anticipated wave energy for that particular location.

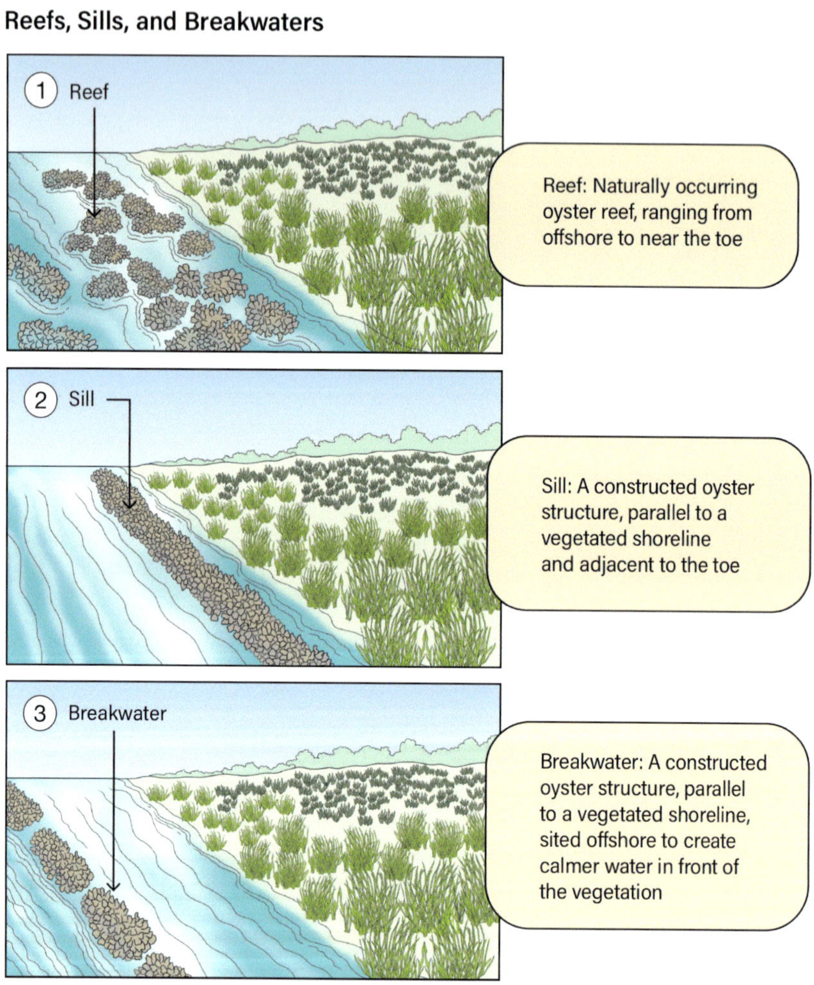


**FIGURE 5.5** The difference between oysters in reefs, sills, and breakwaters. *(1)* Oyster reefs are naturally occurring formations, and they can be found offshore or near the shoreline and the toe. *(2)* Sills can use oyster structures constructed adjacent to the toe, parallel to a vegetated shoreline. *(3)* Breakwaters using oysters are similar, except they are sited somewhat offshore to create an area of calmer water next to a vegetated shoreline. The "toe" is near the mean low water (MLW) line and is the deepest region where marsh plants begin to grow. *Illustration by Marjorie Shropshire.*

There are many options for oyster reef, sill, and breakwater materials. See below for details about options that are available and allowable at the time of this writing. However, new materials are entering the market and others are leaving the market all the time. In addition, regulations are subject to change. Because of these changes, it is important to work with a permitting official and/or restoration professional to ensure specific materials are available and allowable before deciding on one particular option. Also, consider the carbon footprint of the materials used in the living shoreline. Heavy materials that are produced or sourced from far away will come along with a significant transportation-related carbon footprint. In addition, there are high carbon emissions associated with the production of portland cement, which adds to the carbon footprint of many of the prefabricated options.

## Bulk Materials

### Natural Rocks (Limestone, Granite, or Coquina)

Natural rocks are a popular choice for building reefs and breakwaters due to their natural durability, availability, and suitability for marine environments. When selecting the most appropriate rock type for a shoreline, consider factors such as regional appropriateness, availability, rock size (diameter), quantity, delivery, and placement procedure. In Florida, limestone and coquina are more regionally appropriate than granite. Granite rocks must be imported from Georgia or even farther away, as there is no natural source of granite rock in Florida. This leads to increases in cost and carbon footprint. However, in other states such as Maryland and Virginia, granite sills and breakwaters are common. Limestone rocks can be purchased from local quarries or regional rock suppliers. Nearby quarries may offer the best option as they may provide bulk quantities of rock with a lower overhead cost and smaller carbon footprint due to the shorter transportation distances required. If there is not a nearby quarry or if nearby quarries do not sell directly to the public, regional rock supply companies often stock limestone and coquina rocks in various sizes, making them accessible options for smaller projects. However, regional suppliers often have higher prices. Some online marketplaces may offer limestone or coquina rock for sale. Be cautious about shipping costs and verify the seller's credibility before opting for an online supplier.

Generally, larger and heavier rocks tend to be more stable in coastal environments, but larger rocks are also more expensive and more difficult to handle. The size of rocks can be categorized using the value D50, which represents the median diameter. D50 refers to the diameter (D) of the fiftieth percentile (median) of any given set of rocks. For example, a D50 value of 8 inches means that some rocks will be smaller than 8 inches in diameter and some will be larger, but the median size of that set of rocks is 8 inches in diameter. This classification is used because there will always be some variability in rock sizes due to the quarrying process. Smaller rocks with a D50 value of, say, 6 inches will be more easily moved by waves compared to larger rocks with a D50 value of 18 inches. Sometimes, smaller rocks (filler stones) are used to build the core of a structure and are then covered by a layer of larger stones (armor stones). Smaller rocks can also be used to fill a gabion framework. A professional can help to determine the appropriate size and configuration of rock structures. It is highly recommended that a professional be consulted, especially for higher-energy sites.

Calculate the required quantity of rocks needed based on the dimensions and scale of the reef or breakwater project. Most rock suppliers price and sell materials by cubic yards or tons, so keep this in mind when calculating the quantity to order. It is advisable to order an extra 5 to 10 percent to account for any unexpected variations in rock size, breakages during shipping or placement, and settling of the reef structure after placement. If the rocks are not readily available from a nearby location, arrange transportation from the quarry or supplier to the site location.

For large-scale projects, heavy machinery and barge-like vessels with cranes may be required to move and place the rocks. For smaller-scale projects, limestone rocks can be loaded onto small floating platforms or into carts and placed by hand in small batches in the reef footprint.

## Recycled Clean Concrete

Using clean, recycled concrete for building reefs or breakwaters is a beneficial way to repurpose construction waste for marine conservation and coastal protection. The process of recycling concrete involves breaking up and reusing old concrete structures, slabs, or debris. Using clean recycled concrete for reefs and breakwaters not only reduces construction waste but also has a lower carbon footprint than purchasing new concrete substrates. However, it is crucial to ensure reputable sourcing because the concrete must undergo a cleaning process to remove any contaminants or harmful substances before it is used for marine construction. Reliable and reputable concrete recycling facilities follow strict quality-control measures to ensure its durability and performance in the marine environment.

The concrete recycling facility will process the old concrete into specific sizes suitable for each project. The resulting recycled concrete aggregate (RCA) should be free from excess dust, debris, or foreign materials. The process of building a reef or breakwater with clean recycled concrete is like using natural limestone rocks. Calculate the amount of material needed in cubic yards and consider purchasing a marginal amount of extra material. Placement methods for clean concrete are similar to those for natural rock described above.

## Prefabricated Concrete Materials

### REEF BALLS AND LAYER CAKES

Reef Ball is a trademark of Eternal Reefs, LLC. Reef Balls are patented, dome-shaped artificial structures that have been used in marine habitat restoration and enhancement projects for several decades. Reef Balls have been used around the world in projects in close to seventy countries. They are made of marine-grade concrete that is poured into a mold to create a hollow, dome-like structure with many openings and a rough, textured surface meant to mimic the natural characteristics of a coral reef or oyster reef. The hollow interior provides space for marine organisms to live and seek shelter, and strategically placed openings facilitate water flow and create additional hiding places for small fish and other marine creatures. The exterior surface of Reef Balls has a rough texture that creates a favorable substrate for marine organisms. Reef Balls come in various sizes (e.g., oyster ball or goliath ball) to suit different restoration and habitat enhancement objectives. The project conditions, such as depth and tidal range, will determine the correct Reef Ball size. Small sizes of Reef Balls can be placed by hand. Medium-sized balls can be placed using team carry using a metal bar passed through the holes to facilitate lifting. Larger sizes of balls must be placed using a barge and crane and/or float bladders. Many contractors are equipped to place Reef Balls should the homeowner need assistance. The company also offers a product called Oyster Layer Cake that comes in all the same sizes as Reef Balls. Layer Cakes are modified so that the dome is formed out of a series of stacked discs with space in between. These create a different kind of structure than the hollow dome of the standard Reef Ball. See the Reef Innovations website (https://reefinnovations.com/).

### WAVE ATTENUATION DEVICES (WADS)

Wave Attenuation Device (WAD) is a trademark of Living Shoreline Solutions Inc. WADs are patented, pyramid-like structures designed to dissipate wave energy using a system of strategically engineered holes and shapes. They have been used in coastal projects for more than twenty years and are best suited for large-scale and higher-energy applications. Due to their large size, professionals are needed to design the project and supervise the deployment with heavy equipment to ensure the specific orientation and other design parameters are met for each location's conditions. More recently, Living Shoreline Solutions has also created smaller WAD modules for use on moderate- to low-energy shorelines, and ECO WADs with more complex surfaces that encourage more marine life colonization. See Living Shoreline Solutions' website (https://livingshorelinesolutions.com/).

### OYSTER CASTLES

Oyster Castle is a trademark of Chaney Enterprises. Oyster Castles are similar in appearance to cinder blocks but are modified into castle-like shapes that allow blocks to interlock and stack. Oyster Castle blocks are designed as a modular system, allowing them to be easily stacked or interlocked to form larger structures, such as reefs or breakwaters. This modular design provides flexibility and adaptability, enabling various configurations to suit different restoration projects. The surface of Oyster Castle blocks is textured to create an ideal substrate onto which oyster larvae can settle and attach, and they are made from a proprietary blend of concrete. There is an optional formulation that includes crushed oyster shells within the aggregate used to produce the block. The rough surfaces provide attachment points for the oyster spat (free-floating, young oysters) to secure themselves, enabling them to grow and develop into mature oysters. Oyster Castles are produced mostly in Virginia, so be sure to consider transport costs and carbon footprint when opting for this material for use in Florida projects. (See Chaney Enterprises, https://www.chaneyenterprises.com/buildingmaterials.)

### REEF ARCHES

Reef Arches is a registered trademark of the Reef Arches company and Hedrick Brothers Construction. Reef Arches are patented arch-shaped modules made from calcium sulfoaluminate (CSA) concrete with an internal honeycomb-like structure. (See below, under Marine-Friendly Concrete, for more.) Modules are available in different sizes, such as the smaller 50-pound (23-kg) arch and the larger 1,200-pound (544-kg) arch. Units can be connected for added stability and stacked into different configurations. In addition, inserts are available for larger modules that nest into the hexagonal honeycomb cells to create added surface area for marine life attachment. Reef Arches can be used to attenuate wave energy and enhance both oyster and mangrove habitat, depending on the water depth where the units are installed. Find out more at https://www.reefarches.com/.

### QUICKREEF

QuickReef is a trademark of Native Shorelines, a Davey Company. QuickReef is a patented system consisting of a central spine that creates vertical relief along which modular panels are stacked, usually on both sides of the central spine.

QuickReef panels are made from natural materials like limestone marl and oyster shells, and their complex surface provides an ideal substrate for oyster recruitment while the vertical relief and roughness also dissipate waves. QuickReef systems can be installed as a sill or in a breakwater configuration, and they can be adapted for higher energy environments. Suitable for low- to moderate-energy sites, they enhance shoreline stability even in areas without native oyster populations. (See Native Shorelines' "QuickReef" page, https://nativeshorelines.com/living-shoreline-materials/quickreef/.)

## Marine-Friendly Concrete

Regardless of the module structure, concrete formulations that include environmentally friendly cement should be used. Environmentally friendly concrete is benign to marine life, and organisms such as oysters and mussels will attach to its surfaces. An environmentally friendly cement and concrete mix is also better for the planet because it uses materials that have lower carbon emissions. Concrete production accounts for 8 percent of total global carbon emissions and 2.8 billion tons of carbon dioxide annually.[5] Efforts to make "greener" concrete include the use of cement materials that emit lower amounts of carbon by replacing the percentage of commonly used portland cement in the concrete mix, such as fly ash, slag (a byproduct of iron production), and other recycled materials.[6] Portland cement, the most widely used building material worldwide, accounts for 5 percent of man-made carbon emissions.[7] In addition to reducing the proportions of portland cement, other lower carbon-emitting cements such as calcium sulfoaluminate (CSA) can replace portland cement entirely. For oyster prisms and volcanoes (described below), CSA cement was chosen for its low carbon emissions, fast-setting, and increased calcium content, which is more favorable for oyster shell attachment. The acidity of the cement used in the concrete mix is also an important characteristic that makes it marine-friendly. Typically, concrete is highly alkaline (basic) with pH values greater than 11. Just for reference, the pH of seawater is 8, and lower salinity water in estuaries will often have pH levels below 8. Most freshwater has a more neutral pH between 6.5 and 8. The development of marine-friendly concrete mixtures can result in pH values of 9 to 10, lower than portland cement, making them more attractive to settling organisms than traditional portland cement.[8] Other companies use a concrete mix (calcium carbonate [$CaCO_3$] enriched) favorable for oyster settlement in their modules (e.g., Oyster Castles). One company even sells the concrete mix as a separate product (see, e.g., Grow Oyster Reefs LLC's patented Grow Concrete Mix, https://www.growoysterreefs.com/). Although these formulations may be proprietary, a restoration practitioner can experiment by using a lower pH cement such as CSA and possibly adding additional materials that are marine-friendly, such as crushed shell.

The range of reef modules that can be used for hybrid living shorelines, and the materials used in creating them, as described above, is just scratching the surface. Continual research and development in this area are leading to advances in making more environmentally friendly concrete formulations and physical designs that dampen waves and currents and that may even be aesthetically pleasing. Newer manufacturing technology such as 3-D printing can expand the range of designs that can simulate natural forms like wild oyster reefs, at a range of production scales for both small community groups and large private firms.[9]

## Oyster Shell–Based Materials

### SHELL BAGS

Shell bags, also known as oyster bags or oyster shell bags, are some of the most common materials used to create reefs in living shorelines. These bags are typically made of durable mesh material and are filled with discarded oyster shells collected from seafood restaurants, oyster farms, or other sources. Many areas of Florida have shells or even finished bag units available through local recycling programs. The shells are best for providing a substrate for new oyster larvae to settle in and grow on. The collected oyster shells are first cleaned and sorted to remove any debris or organic material and then cured in the sun for several months. Then, 20 to 30 pounds (9 to 14 kg) of shell is placed into a mesh bag that is typically 2 to 3 feet (0.6 to 0.9 m) in length and 8 to 12 inches (0.2 to 0.3 m) high. Shell bags can be constructed and filled by hand and are often made in bulk by volunteers working with nonprofits and other groups around Florida.

The mesh bags allow water to flow through while keeping the shells contained; unfortunately, the mesh is also plastic (usually Naltex Duronet is used). With science now showing the pervasiveness of plastics in our oceans, the bagged shell technique has experienced increased scrutiny for the potential negative impacts of using plastic mesh for oyster restoration. State regulatory agencies now promote non-plastic materials and actively discourage plastic use in restoration. While these potential impacts are only beginning to be explored, there have been concerted efforts within the restoration community to move away from using plastic. More eco-friendly materials to create a wide diversity of reef modules are now being developed. Many of these new techniques, such as mesh bags made from natural basalt fibers, show great promise in creating viable oyster habitats without introducing plastic to the marine environment. Living shoreline contractors and others should be versed in these new methods and can determine the most effective approach to constructing oyster habitat as part of the living shoreline project. But, at the time of this writing, traditional shell bags are still widely available and commonly used.

The filled shell bags are usually placed by hand and stacked multiple layers high, with a wider base layer for stability. Alternating bag orientations are used on each layer for added stability. The height of the shell bag stacks is determined based on the optimal oyster elevation for each site. The goal is for larvae to attach themselves to the oyster shells, to grow and develop into young oysters, and eventually to cement together the shell bags into a contiguous reef.

In areas where this is successful, the plastic mesh is less concerning as it is buried below layers of living oysters. However, if the reef fails to establish, shell bags can become a liability because the mesh degrades and breaks apart when exposed to sunlight, saltwater, and wave action. Once mesh bags begin to break apart, they will need to be removed as this material can pose a serious entanglement risk to wildlife. Therefore, shell bags are best applied in locations that have high oyster recruitment and survival rates.

### SHELL MATS

Shell mats are another material that relies on plastic mesh to contain oyster shells. These mats, developed at the University of Central Florida, are produced by attaching oyster shells with a hole drilled at the base to a flat piece of Vexar mesh (about 2.7 square feet [0.3 $m^2$]) using plastic cable ties. The mats are then lined up and weighed down by attaching the corners of each mat to cement rings. The cable ties hold the oyster shells upright, creating a higher level of vertical relief. Shell mats are often produced and deployed by volunteers. They have been used with success to restore low-profile oyster reefs, mostly in Florida's Indian River Lagoon. Recently, biodegradable materials have been used in oyster mats, enabling the use of this successful method without plastics.[10]

## Plastic-Free Alternative Materials

Shell bags are widely used in living shorelines because they are inexpensive to produce, easy to build and deploy using volunteer assistance, offer a high degree of flexibility in project design (i.e., they can be stacked in any configuration), and they have a history of success in many states. Because of these factors and the relative lack of viable plastic-free alternatives that encompass these benefits, the living shorelines community has been slow to move away from shell bags and other plastic-based materials. Perhaps the desire to continue working with volunteers and communities to build living shorelines is an overarching reason why shell bags remain popular. Even so, there is a large movement, especially in Florida, to find ways to eliminate plastics from living shorelines.[11] One benefit of biodegradable materials is that if they fail, removing them will probably not be necessary (in contrast to shell bags) because the materials generally do not pose a risk to wildlife or other ecosystems.

Several of the options previously discussed such as domes, blocks, and rocks are plastic free. While these alternatives may work in many contexts, most of them cannot be produced locally or by volunteers, they can be hard to deploy using volunteer labor, and they may cost more than shell bags or are not readily available in all markets. Therefore, shell bags often prevail over these alternatives.

Many of these alternatives have their own environmental downsides, such as the high carbon footprint associated with portland cement and long-distance transport of heavy cement or rock from factory or quarry locations. Therefore, much attention has been focused in recent years on developing alternative materials that can be produced using hand labor (usually relying on volunteers) and non-proprietary materials, and that are modular and lightweight enough to be deployed by hand, without special equipment.[12] Carefully consider the trade-offs for the project when selecting a material. Many plastic-free alternatives are new materials that are still undergoing testing to determine what contexts are most suitable for their use.

### GABION CAGES

Gabion cages or gabion baskets are similar in many ways to shell bags but rely on wire mesh rather than plastic mesh to contain oyster shells, rocks, or other materials. They can be rectangular, cylindrical, or prism-shaped and are usually made of galvanized or plastic-coated metal wire mesh (though plastic-coatings should be avoided whenever possible). They are modular and relatively cost-effective for oyster reef creation. Gabions can be constructed in any dimensions to fit the needs of the project, though many gabion suppliers offer a set of standard sizes. They can also be used in upland applications such as for terracing when filled with rocks or bagged sand or soil. Oyster gabion cages are relatively easy to handle and transport (unless very large sizes are used), making them convenient for oyster restoration projects. They can be constructed and filled off-site and deployed later, or they can be constructed and filled on-site to reduce the need to transport and carry fully assembled units.

The wire mesh of gabions can degrade over time as the metal oxidizes (rusts) and this comes with associated concerns about the release of heavy metals into the environment. If oysters are successful at the site, an established reef will be left behind even if the cages rust away. Check on the gabions frequently after deployment for signs of live oyster cover and reef establishment. If the gabions are degrading but the oyster reef is not fully established, additional gabions will be needed, maybe with thicker gauge wire.

A project in Florida's Guana-Tolomato-Matanzas National Estuarine Research Reserve reported good outcomes using gabions with 1 × 1 inch (0.25 × 0.25 cm) mesh and 12.5-gauge galvanized after welded (GAW) mesh.[13] Likewise, recent gabion-based projects in the Indian River Lagoon show promise on local shorelines.[14] If a material closer to shell bags is desired, using thinner gauge metal mesh creates soft-sided gabions that can be stacked and transported more like plastic mesh shell bags. The trade-off with thinner mesh is faster rusting and degradation. Note that there are also materials such as ReefBlkSM that are large-scale, prefabricated gabion structures that are designed for reef and breakwater applications. (See more on ReefBlkSM at Coastal Environments Inc., https://www.coastalenv.com/vertical-oyster-reef-configurations.)

### PERVIOUS OYSTER SHELL HABITAT (POSH) SUBSTRATES

POSH substrates are clumps of oyster shells bound together using cement.[15] They were developed at the University of North Florida and are designed to mimic the natural oyster reef environment where adult oysters are bound together and provide a stable, complex surface for oyster larvae attachment and reef formation. These substrates are designed to allow oyster larvae to settle in various microhabitats within the structure, promoting increased settlement rates and young oyster survival. The permeable nature of POSH substrates allows water to flow through the structure more freely. This enhanced water flow aids in delivering nutrients and oxygen to the oysters and helps flush away sediment and waste, potentially maintaining better water quality within the restoration site. POSH substrates are engineered to be durable and long-lasting, in contrast to some of the biodegradable, metal, or plastic materials. Use of a more durable substrate increases the chance that the oyster reefs created by the substrates can persist over time, which contributes to the long-term success of reefs. This is especially important in areas that may experience periodic oyster die-offs due to freshwater events or seasonal stress related to water level changes, as observed in the Indian River Lagoon. POSH substrates are a relatively new material, but early tests in Florida indicate successful oyster recruitment and establishment on the substrates.[16] At present, POSH substrates are not commercially available, but they are simple to construct and can be produced at home by following methods outlined in the POSH manual.[17]

**FIGURE 5.6** Some of the materials options available for building reefs and sills for living shorelines: *(A)* Natural rock; *(B)* Reef Balls; *(C)* Wave Attenuation Devices (WADs); *(D)* Oyster Castle blocks; *(E)* traditional plastic mesh shell bags; *(F)* non-plastic shell mats; *(G)* gabion cages; *(H)* Reef Prisms; *(I)* Oyster Catcher substrates; *(J)* Community Oyster Reef Enhancement (CORE) modules; *(K)* Biodegradable Ecosystem Engineering (BESE) elements; *(L)* cement-coated jute; *(M)* Natrx basalt bags; *(N)* QuickReef; and *(O)* Reef Arches. *Photos A, D, E, and I by Savanna Barry. Photos B and H, UF/IFAS file photos by Tyler Jones. Photo C by Sydney Walsh/Audubon Florida. Photo F by L. Walters. Photo G by GTM NERR. Photo J by UF Whitney Laboratory. Photo K by Sanibel-Captiva Conservation Foundation (SCCF). Photos L and M by L. Walters. Photo N by The Davey Tree Expert Company. Photo O by the Marine Resources Council at the Ted Moorehead Lagoon House.*

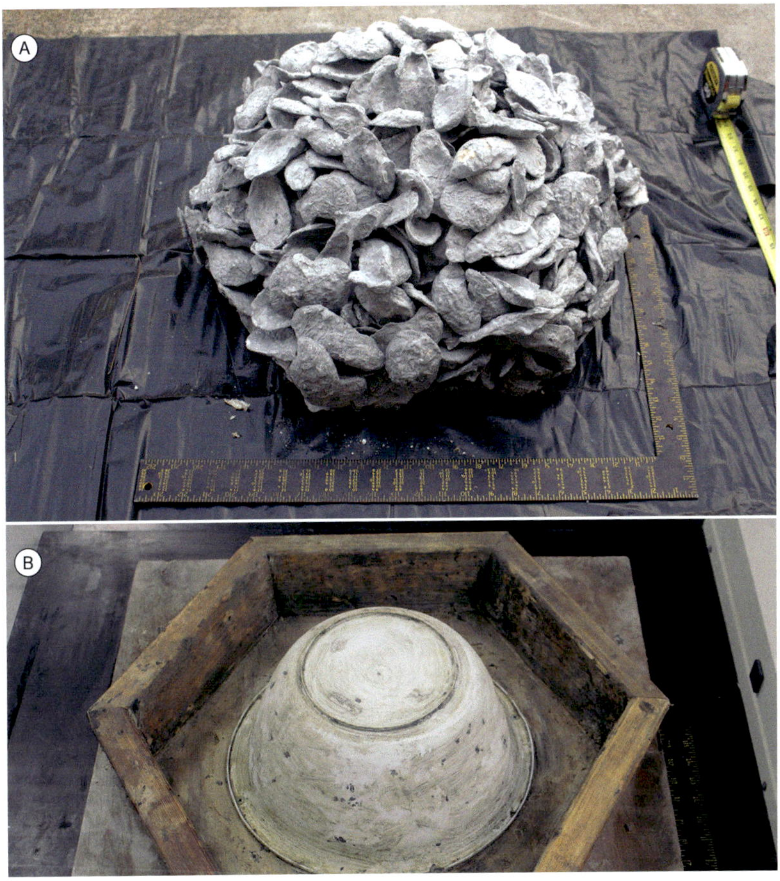


**FIGURE 5.7** POSH modules: *(A)* image of a completed POSH module, and *(B)* the mold used to construct the module. *Photos by Mohammad J. Uddin.*

### SHELL CORRALS

This approach mixes rock and shell materials in order to take advantage of the benefits of both. Durable and heavy stone (usually limestone in Florida) material is stacked around the outer perimeter of the reef footprint to form a "corral," or fence, that is then filled with loose oyster shells. The aim is to create a more complex and natural substrate using oyster shells on the interior, while relying on the larger stones to contain the loose shell and increase stability without the need for plastic bags or other containment mechanisms. A twist on this approach is to build the outer corral using gabion cages or shell bagged in coir fiber bags rather than rock.[18] Refer to the sections above about natural rocks, gabions, and oyster shell bags for more information about material selection and sourcing.

### REEF PRISMS AND VOLCANOES

A development team at the University of Florida spearheaded the creation of reef prisms. Reef prisms (and other shape varieties) are made from a combination of jute-fiber erosion control mat and a rapid setting, low-shrink grout (CSA) available from local big-box stores. (The chemical formulation of CSA is calcium sulfoaluminate; its commercial name is Cement-All.) CSA has a higher break strength, a 35–50 percent lower carbon footprint, a more neutral pH favorable for marine organisms (8.5 vs. 12.1), and a significantly faster setting time compared to portland cement. The lower carbon footprint results from lower energy inputs during the cement material production. Reef prisms absorb, rather than simply deflect, waves due to their shape and semi-porous construction. Their triangular dimensions can be adjusted based on site conditions, modules can be easily stacked, and placement can be tailored to a wide variety of reef and shoreline designs. There are also many shape variations that can be created from the jute and CSA material, including ribbed panels, half-pipes, domes, and even inverted nest-like domes dubbed "oyster volcanoes."[19] Oyster volcanoes were inspired by reef prisms and pioneered by practitioners in the Indian River Lagoon region of Florida. The process for making reef prisms is completely open-sourced and available through the University of Florida/Florida Sea Grant program. Briefly, jute is rolled out, cut to size, coated with the grout mixture, spread out onto a flat surface, and wrapped around a triangular form (the standard size is 1 ft. tall × 4 ft. long [0.3 m tall × 1.2 m long]). Once the material cures, the form is collapsed and removed. Reef prisms are usually filled with oyster shells for added weight and habitat complexity. Find out more at the PROS, Plastic-Free Restoration of Oyster Shorelines website (https://sites.google.com/ufl.edu/pros/home).

At the time of this writing, reef prisms and other shapes are undergoing testing and refinement through pilot projects run by an interactive team of practitioners. There are several pilot locations where prisms have been deployed and monitored over time. Reef prisms are a relatively new material, and these recent pilot tests have so far shown they perform best where oyster recruitment and survival is high. As with gabion cages and shell bags, the goal is for oysters to recruit onto the substrate, bind loose shells together, and form a self-sustaining reef, even if the base material (metal, coated fiber, plastic) degrades. As reef prisms biodegrade, the oyster shell within them will be left behind, along with any live oysters coating that shell. However, in places where oysters do not attach and coat the substrate, prisms may degrade too quickly for an oyster reef to establish. Nonetheless, even with limited oyster recruitment, prisms formed into breakwater reefs have shown significant sediment accretion landward of the structures and increased sediment stability, and when used as a sill, prisms have protected planted vegetation. Reefs made of this material in Cedar Key, Florida, where oyster recruitment is high, have persisted since 2018 (more than five years at the time of this writing) while prisms have degraded in less than a year at some other pilot locations with low oyster recruitment.[20] Overall, reef prisms are best suited to lower-energy shorelines with high oyster recruitment and stable salinity (i.e., not areas that commonly get freshwater events that cause oyster die offs). They may be particularly suitable as a material for sills (protective edges for vegetation), similar to coir logs, or as temporary breakwaters to accumulate sand into which vegetation can later be installed.

### OYSTER CATCHER SUBSTRATE

Oyster Catcher is a trademark of the Sandbar Oyster Company. Oyster Catcher substrates are made from natural fibers coated in mineral hardener (cement) and woven or shaped into many forms, including tables, lollipops, lattices, pillows, and rods. These forms can be combined and adapted into custom designs to suit the needs of the site. Oyster Catcher substrates may be particularly valuable in areas with soft, muddy bottoms, as the rods and other vertical structures can hold the reefs off the seabed, preventing their smothering by sedimentation or the settling of substrates into the mud. Oyster Catcher substrates have been deployed in several locations in the Southeast, including North Carolina, Georgia, and Florida's Panhandle and Indian River Lagoon. Studies have reported successful outcomes for oyster recruitment and nekton (mobile invertebrates and fish) use at reefs constructed from Oyster Catcher substrates as well as some success for marsh protection.[21] Similar to reef prisms and other biodegradable substrates, they are most appropriate for use in locations where a self-sustaining oyster reef is expected to develop over time, as the substrates will eventually biodegrade. Use of Oyster Catcher substrates requires licensing and training from Sandbar Oyster Company. See the company's website (https://www.sandbaroystercompany.com/).

### COMMUNITY OYSTER REEF ENHANCEMENT (CORE) MODULES

CORE modules were developed by a team at the University of Florida's Whitney Lab as a plastic-free alternative to shell mats. CORE modules are lightweight and are easy to produce and deploy in rectangular arrays, similar to shell mats. They are cast in shallow pans using a blend of sawdust, burlap, and portland cement. Before the block sets, oyster shells are placed in vertical orientation to enhance structural relief/complexity and create microhabitats for oyster settlement and other reef organisms. CORE modules are typically produced by volunteers and have been used most commonly in northeast Florida. Find out more about CORE modules at https://www.whitney.ufl.edu/conservation--sea-turtle-hospital/oyster-restoration-initiative/.

#### BIODEGRADABLE ECOSYSTEM ENGINEERING (BESE) ELEMENTS

BESE Elements is a trademark of BESE Solutions. BESE (pronounced "beez") elements are three-dimensional structures constructed of a bioplastic (PLA) material derived from potato starch, a byproduct of potato chip production. They have been used across the globe in a variety of habitat restoration projects such as stream banks, seagrasses, and oyster reefs. BESE Elements are modular and can be used in several ways for living shorelines. They can provide a matrix for oyster attachment, or they can be used to stabilize new plantings of mangroves or marsh grasses. BESE Elements are fully biodegradable, and the rate at which they biodegrade depends on ambient conditions such as temperature, nutrient supply, and wave energy. The company estimates a lifespan of two to four years in most tropical locations and between ten and twenty years in temperate locations. BESE material also comes in mesh and mat formations and can be used to produce plastic-free versions of shell bags and shell mats. The plastic-free version of shell mats, also developed at the University of Central Florida, rely on two layers of BESE material in place of the plastic mat and on stainless-steel twist ties instead of zip ties to attach the oyster shells to the BESE mat. Field measurements indicate the BESE-based mats performed as well as the plastic oyster mats in terms of oyster recruitment across two years of study.[22] Based on measurements of mass loss over time, researchers estimate that BESE-based mats in the Indian River Lagoon will have a half-life between 4.4 and 6.7 years (which translates to a total lifespan of 8.8 to 13.4 years). A study in Texas found that BESE-based mesh bags degraded within two months but that oysters were still able to grow on the loose shell left behind.[23] See the company's website to learn more about BESE products (https://www.bese-products.com/).

#### BUILDING BLOCKS FOR EDGING AND SOIL STABILIZATION

Depending on the bank slope, wave climate, soil type on the bank and nearshore, additional stabilizing elements may be required. The materials below are examples of both temporary and long-lasting materials that can be used either to underlay or to support the main vegetation and oyster reef zones of a living shoreline.

### Fiber Logs

Standard fiber logs are large rolls 10 feet (3m) long and 10 to 20 inches (25 to 51 cm) in diameter. Made of woven fiber, these "logs" (made of jute, coconut fiber [coir], or other natural fiber) can be placed along the shoreline to hold sediment in place and support new plantings. (See chapter 4 for more about living shoreline designs that include temporary edging.) Fiber logs can be used to create terraces or to support the waterward edge of low marsh plantings during initial establishment. Manufacturers estimate that fiber logs last two to five years (sometimes up to ten years), but in saltwater applications in Florida's warm climate, many logs biodegrade more rapidly (six months to two years). It is critical to dig a trench one-half to two-thirds of the log's diameter to seat the logs into the sediment. Additionally, once the logs are in place, they must be anchored to prevent them from being moved or washed away by waves or currents. In most cases, wooden stakes placed every 2 to 3 feet (0.6 to 0.9 m) are sufficient to anchor the logs in place. In addition, plants are often inserted directly into the logs and provide even more anchorage. Fiber logs are most useful in lower-energy settings where plants will provide sufficient wave attenuation after they have established themselves. If longer term energy protection is required to maintain plant survival, a sill made from more durable materials should be considered.

## Geotextiles

A geotextile is a permeable textile or fabric intended to stabilize soils, facilitate drainage, and/or serve as a base layer for landscaping elements or larger infrastructure projects. Common natural fibers formed into geotextiles include burlap, coir, and jute. These are sold in large rolls and are often marketed as erosion control mats. If a shoreline has a steep bank, the project may benefit from a geotextile layer. Geotextiles also come in synthetic (plastic-based) forms, and these are more often used as a base layer for rock revetments. Geotextiles come in various forms, including woven and nonwoven types, and they are selected based on the specific engineering requirements of a project. The choice of geotextile depends on factors such as strength, permeability, and the intended application. While there are some bank-stabilization applications where synthetic geotextiles may be recommended, natural fibers should be used whenever possible. Although often sold as weed-barrier cloth, it is not recommended for this purpose in Florida. See chapter 9 for more details on this.

## Geogrids / Geowebs

Geogrids are a synthetic material used in shoreline stabilization projects to reinforce and stabilize the soil and prevent erosion, especially when the bank slope is steep. They are typically made of high-strength plastic polymer materials, such as polypropylene or polyester, and they feature an expandable grid-like structure. In living shoreline projects, geogrids can be used to establish plants by holding soil in place. The grid sheets are expanded along the bank and filled with soil, and then plantings are added to the grid cells. It is often necessary to plan for long-term maintenance of geogrid structures, including filling in cells where sediment may get washed out over time, especially when installed on a slope. In addition, if the sediment in the area where reefs are planned is very soft or if scouring around reef structures is a concern, a geogrid mattress can be placed under the reef area to prevent scouring and settling. Use of geoweb and geogrid materials should be considered carefully, as it is common for these materials to become undermined and fail catastrophically when improperly installed, poorly maintained, or incorrectly sited, necessitating complete removal and replacement.

### Envirolok Bags

Envirolok is a trademark of Agrecol, LLC. Envirolok is a proprietary system designed for erosion control and vegetative retaining walls. Essentially, it is a bag-based system that uses soil and plants to create mechanically stable structures. The Envirolok system consists of bags made from synthetic geotextile fabric that does not biodegrade. These bags are filled with soil, compost, or other organic material and then stacked in a way to create a retaining wall or erosion-control structure. Once the structure is in place, vegetation can be planted into the bags, and as the plants grow, their roots intermingle with the soil inside the bags, creating a natural, reinforced system. Over time, the roots of the planted vegetation grow through the fabric, intertwining and forming a robust network that reinforces the barrier against the forces of waves and tides. The integration of vegetation not only helps stabilize the structure but also provides habitat and promotes a more natural appearance, making it a popular choice for certain ecological restoration and landscaping projects. Learn more at Envirolok's website (https://www.envirolok.com/).

### Gabions

Gabion cages, while useful as oyster reef substrates, can also be used in bank-stabilization and terracing applications. (See above for a full description of gabions, page 134.)

### HESCO Barriers

HESCO is a trademark of Hesco Bastion Ltd. HESCO barriers were originally designed for military and security applications but have found a unique purpose in the realm of environmental conservation, particularly in large-scale shoreline restoration projects. These collapsible wire mesh containers, when filled with sand

**FIGURE 5.8** Some of the materials options available for edging and shoreline stabilization for living shorelines: *(A)* Coir logs, *(B)* Geotextile, *(C)* Geoweb, *(D)* Envirolok bags, *(E)* gabion cages, and *(F)* HESCO barriers. *Photo A by Delaware DNREC. Photo B by Ginny Stibolt. Photo C by Savanna Barry. Photo D by Envirolok Geobag System. Photo E courtesy of La fotisto/Shutterstock.com. Photo F by Andy Patton, 2007, Wikimedia Commons (https://commons.wikimedia.org/wiki/File:Porta_john_hescos.JPG).*

or soil, can serve as effective tools to stabilize coastal habitats against erosion. For homeowners living near the coast, using HESCO barriers in combination with native plantings can stabilize banks and safeguard homes while also contributing to the restoration and preservation of natural habitats. These barriers act as a bulwark against aggressive wave action, giving vegetation a chance to grow and hold the soil together, ultimately reinforcing the shoreline's integrity. See more at HESCO's website (https://www.hesco.com/solutions/flood-protection/).

## Building Blocks for Vegetated Zones

Some of the most common living shoreline plants are covered in this section, but there are many more plant options, especially in higher elevations (above the mean higher high-water line [MHHW]) and in lower salinity areas. Detailed information about plant species, including hardiness zones and regional plant lists, can be found in appendix A.

### Mangroves: A Shoreline's Best Friend

The benefits of mangroves are:

- Mangroves act as natural barriers, protecting shorelines from erosion, storm surges, and wave impacts.
- Mangroves support immense biodiversity, including fish, birds, and invertebrates.
- Mangroves are highly efficient at capturing and storing carbon, helping to mitigate climate change.
- Many fish species and crustaceans use mangroves as breeding and nursery grounds, supporting local fisheries.
- Mangrove roots filter pollutants and trap sediments, improving water quality and clarity.
- Mangroves reduce the intensity of floods by absorbing excess water during high tides and heavy rainfall.
- Mangrove ecosystems attract tourists, which in turn supports eco-tourism activities like bird watching, fishing, and kayaking.
- By stabilizing coastlines and reducing the impact of extreme weather events, mangroves enhance the resilience of coastal regions.

Mangroves are remarkable trees and shrubs that thrive along Florida's coastlines, estuaries, and saltwater lagoons, and they provide many benefits for coastal property owners. Unlike most plants, they can tolerate salty water and have developed unique root systems that stick up from the mud; these root systems are used for aeration even when they are submerged during high tide. These roots, often tangled and visible above the water's surface, act as a nursery for numerous fish, crabs, and shrimp, many of which are popular catches for Floridian anglers. Mangroves also play a pivotal role in shielding the coast from erosion, buffering storm surges, and acting as nature's own line of defense. Plus, they are a bird-watcher's delight, attracting various bird species that come to nest and feed among their branches. Further, the tree canopy acts as a windbreak, and the complex root systems stabilize sediment and deter runoff, preventing erosion and preserving water clarity. According to the Nature Conservancy, "Mangroves are most effective in flood risk reduction where they are abundant and located in front of areas with high densities of people and property. In Lee and Collier Counties, closest to where Hurricane Irma made landfall on the Florida mainland, mangroves are estimated to have prevented losses of $802 million and $580 million, respectively. In vulnerable Miami-Dade County, urban coastal mangroves protected high-value coastal properties from over $134 million in potential flood damages."[24] For anyone in Florida, mangroves are not just scenic backdrops; they are crucial protectors and nurturers of the state's rich coastal biodiversity. (See chapter 1 for more details on mangrove ecology, and chapter 7 for more details on maintaining and trimming mangroves.)

Three species of mangroves are found in Florida: red mangrove (*Rhizophora mangle*), white mangrove (*Laguncularia racemosa*), and black mangrove (*Avicennia germinans*). They each occur in specific zones along the shoreline. Red and black mangroves are most often used in living shorelines.

Red mangroves (*Rhizophora mangle*) are adapted to live at the water's edge and are one of nature's best erosion fighters. Black mangroves (*Avicennia germinans*) are the most cold-tolerant species, and they thrive slightly higher along the shoreline than red mangroves. Black mangroves also tolerate higher salt content in soils.

Waterfront homeowners and shoreline managers in Central and South Florida are encouraged to include mangroves as part of living shoreline projects. Start by sourcing red or black mangrove saplings or propagules (seedlings) from local nurseries or environmental organizations, the larger the better in terms of survival rates (but larger planting units may also be more expensive). It is best to plant mangroves during early spring or late summer for the best survival rate.

Generally, mangroves should be planted around the mean high-water line (MHW) and not too much below it. As the mangroves mature, they need minimal maintenance, but they will require protection from human interference, herbicides, and pollutants.

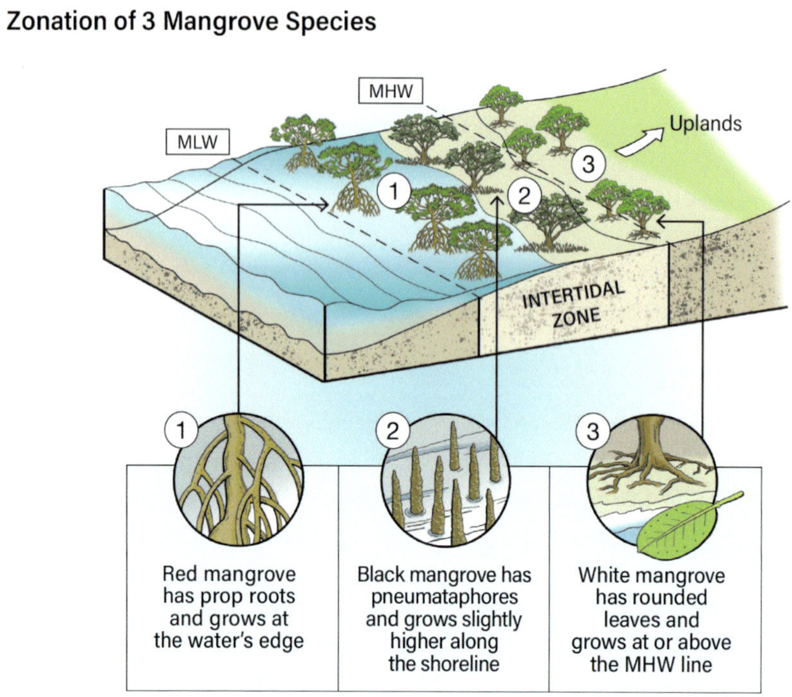


**FIGURE 5.9** Zonation of red, black and white mangroves. Usually, red mangroves grow farthest out into the water, while black and white mangroves grow higher along the shore. The identity of each species can be determined by the shape of their roots and leaves. *Illustration by Marjorie Shropshire.*

Encouraging natural growth around the mangroves, such as seagrasses and other coastal plants, can also enhance the restoration effort. In fact, planting mangroves in conjunction with salt marsh grasses may increase both their success rates.

Mangroves live at the water's edge, so their roots are in the water part of the day. In some cases, they can be planted alone, while on shorelines with more wave energy mangroves can be planted along with oyster reefs. On hybrid shorelines, riprap (natural coquina boulders) can be used as a foundations, or as "planters," for young mangrove trees.

## MANGROVE MYTHS AND MISCONCEPTIONS, INCLUDING MANAGEMENT

Waterfront homeowners and other shoreline managers often have negative views about mangroves and may be reluctant to add them to their waterfronts. This section should help dispel some of the most common myths about mangroves and their management.

Here are some of the misconceptions:

- Mangroves harbor no-see-ums (sand flies), mosquitoes, and other harmful insects.
- Mangroves cannot be managed to accommodate a view.
- Mangroves are messy and cannot be trimmed.
- Mangroves are invasive.

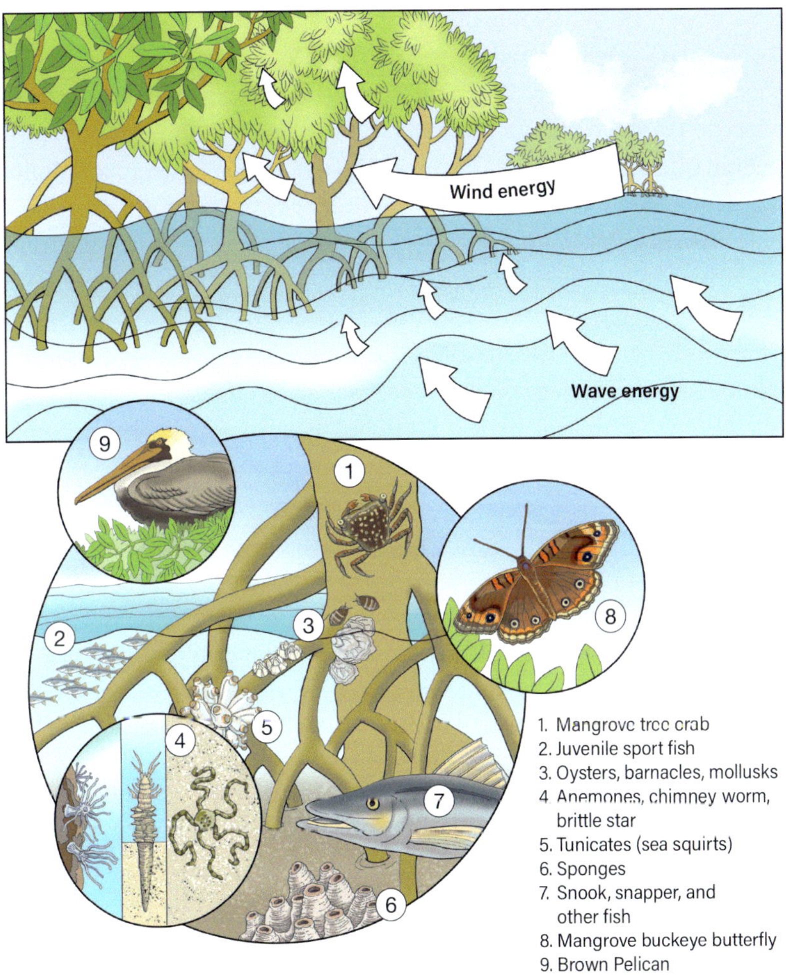


**FIGURE 5.10** Red mangrove roots buffer wave energy, while the tree canopy acts as a windbreak. The tangled prop roots stabilize sediment, prevent runoff from the land, and help preserve water clarity. The entire tree structure, from crown to roots, creates places to forage and hide for a wide variety of creatures. *Illustration by Marjorie Shropshire.*

### MANGROVE ECOSYSTEMS

Contrary to popular belief, planting mangroves does not add to the population of biting insects like no-see-ums or mosquitoes; in fact, mangroves support ecosystems that support predators of mosquitoes and other small insect pests. So, mangroves may reduce biting insect populations.

### MANGROVES AND WATER VIEWS

Mangroves can be trimmed by the homeowner to 6 feet (1.8 m) tall if the tree is not more than 10 feet (3 m) in height. Mangroves grow slowly, so it will take a few years for them to reach the minimum trimming height of 6 feet (1.8 m) tall. Meanwhile, the shoreline will have all the protective benefits they can provide. As long as they are not allowed to grow taller than 10 feet (3 m), homeowners can trim their own mangroves without a permit. For a fact sheet about trimming mangroves, see the Florida Department of Environmental Protection's "Mangrove Trimming Guidelines for Homeowners," at https://floridadep.gov/water/submerged-lands-environmental-resources-coordination/documents/mangrove-trimming-guidelines.

### MANGROVES ARE NOT INVASIVE

Mangroves are native to Florida and so, by definition, they are not invasive. But mangroves do self-propagate if the conditions are favorable. They reproduce by having their seeds germinate while still attached to the parent plant before dropping them as floating propagules. This means that new plants will become established in areas anywhere along the shore just below the mean high tide line. Some people have expressed concern that mangroves have been planting themselves along the edges of established salt marsh areas in Central and North Florida. But this is a natural range expansion process driven by the general warming trend that allows mangroves to out-compete marshes as the freezing temperature events that would normally cause mangroves to die back become less common.[25] (See chapter 2 for more information on invasive plants.)

## Marsh Grasses

The benefits of marshes and marsh grasses are as follows:

- Marsh grasses have dense root systems that stabilize the shoreline and prevent erosion.
- Grasses trap pollutants and sediment, which keeps the water clearer.
- Grasses can create more shoreline by collecting sediment.

- Many species, including commercially important fish and shellfish, rely on marshes as breeding and nursery grounds.
- Marshes are popular for birdwatching, photography, hiking, and educational programs about ecosystems and conservation, supporting the tourism economy.
- Marshes act as natural sponges, absorbing excess rainwater and reducing the risk of floods in surrounding areas.
- Marshes provide critical habitat for a wide variety of wildlife, including birds, amphibians, fish, and insects.
- Like mangroves, marshes capture and store carbon, contributing to the reduction of atmospheric $CO_2$ and combating climate change.
- Coastal marshes help absorb wave energy and storm surges, protecting inland areas from storm damage.
- Many communities have cultural and historical ties to marshland areas, often using them for traditional harvesting of plants and animals.

### HOW MARSH GRASSES ARE USED IN LIVING SHORELINES

There are two species of marsh grasses that are most commonly installed on living shorelines: (1) Smooth cordgrass (*Spartina alterniflora*) is planted in the low marsh area where it is inundated by tides once or twice a day, depending on local tidal patterns. (2) Saltmeadow cordgrass (*Spartina patens*), also called salt marsh hay, is planted above the MHW line. It can tolerate occasional flooding. In the northern part of Florida, cordgrasses can be used together with other salt marsh plants like black needle rush (*Juncus roemerianus*). In southern Central and South Florida, cordgrasses and mangroves are both commonly used in living shorelines.

Smooth cordgrass is the principal species planted on living shorelines in areas with low to moderate wave action. It grows in the intertidal zone and tolerates varying salinities in both estuarine and freshwater environments. The dense root system helps prevent runoff, collects sediment, and has the potential to build shorelines. The stems reduce wave energy, and even a relatively narrow planting can slow wave action significantly. It has been shown that the first 8 feet (2.5 m) of smooth cordgrass can decrease wave energy by 50 percent.[26] At maturity, smooth cordgrass grows to about 1 to 2 feet (0.3 to 0.6 m) above the water, remaining low enough to preserve the view. It tolerates various salinities and can grow in an estuary or in freshwater habitats, though it often gets outcompeted by other species in lower salinity conditions.

**Marsh Grasses at Work**

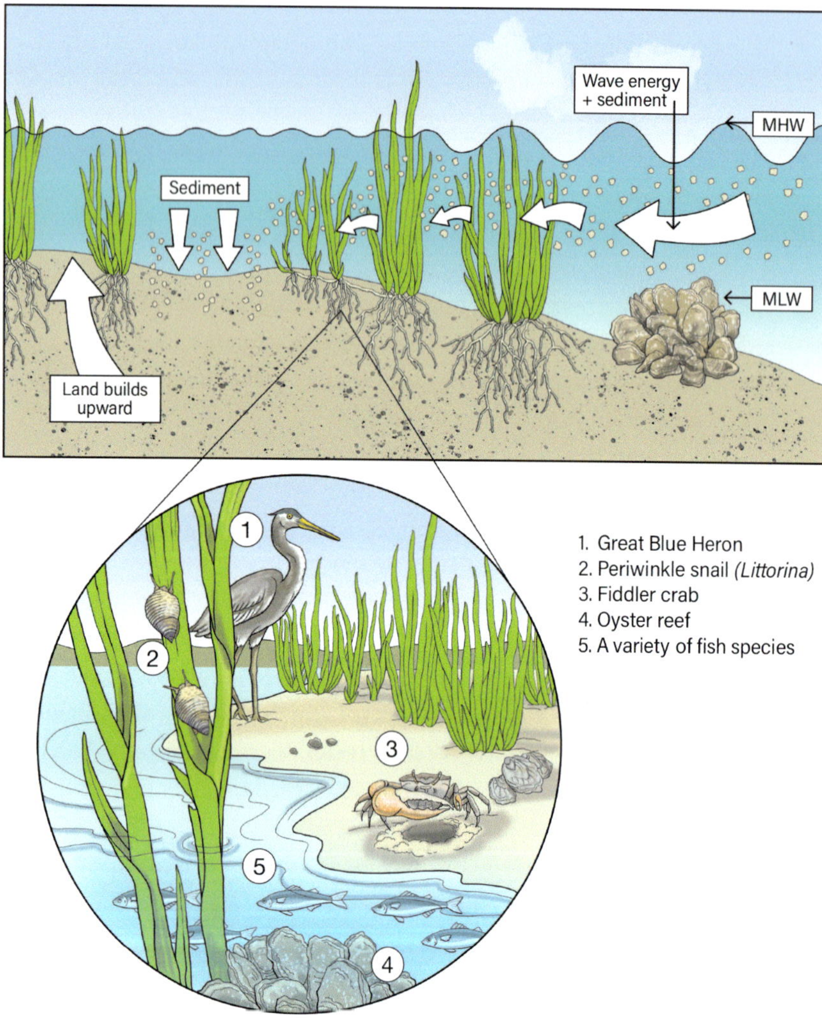


**FIGURE 5.11** Marsh grasses reduce wave energy and capture sediment from the water column, helping to build land. Marsh grasses also provide good habitat for many organisms. *Illustration by Marjorie Shropshire.*

Smooth cordgrass provides good habitat values and offers various creatures places to hide and forage. Organic matter from the breakdown of its leaves provides food for animals low on the food chain, and these, in turn, are prey for other animals. Some of the species commonly seen on living shorelines planted with smooth cordgrass include fiddler crabs, mollusks, pinfish, mullet, and wading birds such as egrets and herons.

**FIGURE 5.12** Grasses in the genus *Spartina* are some of the most ubiquitous species found in natural salt marshes. Therefore, they are also some of the most important building blocks for planted marshes as part of living shorelines. *(A)* Smooth cordgrass (*Spartina alterniflora*) is the only species of marsh grass in Florida that can survive daily inundation by tides (*UF/IFAS file photo by Tyler Jones*). *(B)* Saltmeadow cordgrass (*Spartina patens*) prefers higher elevations and is often a dominant species in the high marsh zone (*photo by Savana Barry*). *(C)* Black needle rush (*Juncus roemerianus*) is commonly found in the transitional zone between the low and high marsh. Black needle rush is especially common in more brackish marshes and along Florida's northern shorelines *(UF/IFAS file photo by Tyler Jones)*.

Saltmeadow cordgrass is like smooth cordgrass in that it can form very dense stands, and it provides many of the same functions in terms of soil trapping and stabilization, pollutant buffering, and habitat provisioning. However, saltmeadow cordgrass does not tolerate daily tidal inundation and therefore thrives at slightly higher elevations (at least one foot [0.3m] above the MHW line), where it will be exposed only to occasional flooding such as might occur during storm events. The nickname salt marsh hay may derive in part from the fact that this grass species was historically harvested and fed to livestock in great quantities, especially in New England states. This grass naturally occurs throughout Florida and is suitable for use on shorelines anywhere in the state.

## High Marsh and Upland Species

A wider diversity of plants is available and recommended for areas above the MHW line. Some of these plants are briefly covered below, but there are many others profiled in the Suggested Plants List provided in appendix A. More information about high marsh and upland landscaping is covered in chapter 8.

Principles of ecological restoration encourage the inclusion of a diversity of plant species whenever possible. Higher plant diversity, assuming the plants are primarily native species, tends to provide a higher level of resistance to disturbance, leads to faster establishment of restoration projects, and provides more varied habitat resources for wildlife. Consider adding some of these plant species to cordgrass-based planting plans whenever possible to enhance diversity along the shoreline.

### GRAMINOIDS: GRASSES AND OTHER GRASS-LIKE PLANTS, INCLUDING RUSHES AND SEDGES

**Bitter panic grass** (*Panicum amarum*): A perennial grass that grows on coastal dunes and sandy beaches. It has a fast-growing, deep root system that helps in stabilizing sand dunes. The grass is characterized by its stiff, blue-green blades and production of airy seed heads.

**Sea oats** (*Uniola paniculata*): A tall, perennial grass with large, feathery seed heads. Sea oats are crucial for dune stabilization in coastal areas. They have long, slender leaves and can grow quite tall: 4 to 8 feet (1.2 to 2.4 m). The plant is highly tolerant of dry conditions, salt spray, and sandy soils.

**Seashore paspalum** (*Paspalum vaginatum*): A spreading perennial grass that is tolerant of salt spray and inundation of brackish or saltwater. It grows 3 to 10 inches (8 to 25 cm) tall.

**Spike grass** (*Distichlis spicata*): A hardy perennial grass species that is extremely salt and drought tolerant. It can grow up to 20 inches (0.5 m) in height and has narrow leaves, each up to 4 inches (10 cm) in length, that grow from the stalks. The plant is dioecious, meaning the male flowers and female flowers grow on separate individuals. The flower stalks may be up to 3 inches (8 cm) long, with green or purple-tinted spikelets.

### FORBS: HERBACEOUS FLOWERING PLANTS THAT ARE NOT GRASS-LIKE

**Blanket flower** (*Gaillardia pulchella*): Blanket flower is a colorful, short-lived perennial known for its bright red and yellow daisy-like flowers. It thrives in well-drained sandy soils and is a popular choice for coastal landscaping due to its drought tolerance and vibrant blooms. It self-seeds. (While this plant is no longer considered to be a Florida native, it is attractive and easy to grow.) Lanceleaf blanket flower (*G. aestivalis*) is native to Central and North Florida and could be an alternative, but it is not salt tolerant.

**Dune sunflower** (*Helianthus debilis*): Dune sunflower is a fast-growing, sprawling perennial known for its bright yellow, sunflower-like blooms. It is a hardy plant that thrives in sandy coastal soils and full sun, making it an excellent choice for beachfront landscaping.

**Glassworts** (*Salicornia* spp.): Glassworts are succulent annuals that grow in salt marshes and muddy shores. They have jointed, green stems and are notable for their lack of distinct leaves. They are often used in saline agriculture and for soil bioengineering due to their salt-tolerance.

**Sea purslane** (*Sesuvium portulacastrum*): Sea purslane is a succulent, sprawling perennial herb that often forms dense mats on sandy shores and salt marshes. It has small, fleshy leaves and produces tiny pink flowers. This plant is salt tolerant and helps in stabilizing sandy soils.

**Seaside goldenrod** (*Solidago sempervirens*): A perennial wildflower, seaside goldenrod is known for its tall spikes of bright yellow flowers. It grows in coastal dunes and marsh edges. This plant is salt tolerant and attracts various pollinators, making it an important component of coastal ecosystems.

### VINES

**Beach morning glory** (*Ipomoea imperati*): This is a fast-growing vine that produces white, trumpet-shaped flowers. Commonly found in coastal sand dunes, it is an important plant for stabilizing sandy soils. The leaves are glossy and ovate.

**Railroad vine** (*Ipomoea pes-caprae*): A fast-growing perennial vine with a spreading habit, railroad vine is known for its large, showy pink or purple flowers. It thrives in sandy, coastal areas and is essential for dune stabilization. Its leaves are fleshy with two lobes, like a goat's foot, the inspiration for its scientific name.

**FIGURE 5.13** Examples of species that thrive in higher elevation zones of shoreline environments can be added to create visual interest and increase plant diversity. *(A)* Sea purslane (*Sesuvium portulacastrum*); *(B)* Glasswort (*Salicornia* sp.); *(C)* Railroad vine (*Ipomoea pes-caprae*); *(D)* Marsh elder (*Iva frutescens*); *(E)* Saltwort (*Batis maritima*); *(F)* Sea oxeye daisy (*Borrichia frutescens*); *(G)* Blanket flower (*Gaillardia pulchella*); *(H)* Bitter panic grass (*Panicum amarum*); *(I)* Sea oats (*Uniola paniculata*); *(J)* Beach morning glory (*Ipomoea imperati*); *(K)* Spike grass (*Distichlis spicata*); *(L)* Seaside goldenrod (*Solidago sempervirens*); and *(M)* Dune sunflower (*Helianthus debilis*). *Photos A, D, I, and K by Savanna Barry. Photo B by Julie Mathis, 2023, Pl@ntNet.org (https://identify.plantnet.org/k-world-flora/observations/1019444413; reproduced with cropping under creative commons license CC BY-SA 4.0. https://creativecommons.org/licenses/by-sa/4.0/). Photo C by 蒹瑜, 2019, Wikimedia Commons (https://commons.wikimedia.org/wiki/File:海邊_馬鞍藤_(Ipomoea_pes-caprae).jpg; reproduced with cropping under creative commons license CC by 4.0. https://creativecommons.org/licenses/by/4.0/). Photo E by Forest and Kim Starr, 2001, Starr Environmental (http://www.starrenvironmental.com/images/image/?q=23902930604; reproduced with cropping under creative commons license CC by 4.0. https://creativecommons.org/licenses/by/4.0/). Photo F by Gail Hampshire, 2017, Flickr*

(caption continued) *(https://flickr.com/photos/43272765@N04/27292382209; reproduced with cropping under creative commons license CC by 2.0. https://creativecommons.org/licenses/by/2.0/). Photo G by John Bradley, 2007, Wikimedia Commons (https://commons.wikimedia.org/wiki/File:Crystal_River_gaillardia02.jpg; reproduced without modification under creative commons license CC BY-SA 3.0). Photos H and M, UF/IFAS file photos by Josh Wickham. Photo J by R. Jones, 2008, Wikimedia Commons, CC BY 2.0. Photo L, UF/IFAS file photo by Joshua Criss.*

TREES AND SHRUBS: WOODY PLANTS

**Marsh elder** (*Iva frutescens*): Marsh elder is a woody shrub found in salt marshes and coastal areas. It has small greenish flowers and toothed leaves. This plant is often a dominant species in high salt marsh environments and provides habitat for various wildlife.

**Saltwort** (*Batis maritima*): Saltwort is a small, salt-tolerant shrub that grows in coastal salt marshes and mangrove swamps. It has fleshy, succulent leaves and produces tiny white flowers. This plant plays a crucial role in salt marsh ecosystems, often as an early colonizer.

**Sea oxeye daisy** (*Borrichia frutescens*): The sea oxeye daisy is a perennial, salt-tolerant shrub commonly found in coastal salt marshes, sandy beaches, and dune areas. It is characterized by its bright yellow, daisy-like flowers and succulent, spoon-shaped leaves. This plant typically grows to about 1 to 3 feet (0.3 to 0.9m) in height and blooms throughout the year, providing a continuous display of color in coastal landscapes and attracting various pollinators.

## Summary

This chapter provided a comprehensive guide to the building blocks for constructing living shorelines, focusing on the use of natural, sustainable, and locally sourced materials. It highlights the significance of choosing materials based on local ecosystems and the ongoing changes in material availability and regulations. The detailed exploration of oyster reefs, breakwaters, alternative materials, and plant species underscores the importance of ecological considerations in shoreline management. This approach not only ensures effective shoreline stabilization and protection but also contributes to the preservation of local biodiversity and the natural environment.

For instance, mangroves are suggested for southern Florida, while herbaceous saltmarsh species are recommended for the northern regions like the Florida Panhandle. Material availability and regulations may vary and change, so consultation with local environmental professionals is encouraged.

## Notes

1. Jennifer E. Manis, Stephanie K. Garvis, Steven M. Jachec, and Linda Walters, "Wave Attenuation Experiments Over Living Shorelines Over Time: A Wave Tank Study to Assess Recreational Boating Pressures," *Journal of Coastal Conservation* 19 (2015):1–11, quoted in L. Walters, M. Donnelly, P. Sacks, and D. Campbell, "Lessons Learned from Living Shoreline Stabilization in Popular Tourist Areas," in *Living Shorelines: The Science and Management of Nature-Based Coastal Protection,* edited by Donna Marie Bilkovic, Molly M. Mitchell, Megan K. La Peyre, and Jason David Toft (CRC Press, 2017), 239.
2. Antonio Rodriguez, F. Joel Fodrie, Justin T. Ridge, et al., "Oyster Reefs Can Outpace Sea-Level Rise," *Nature Climate Change* 4 (2014): 493–97, quoted in C. A. Currin, J. Davis, and A. Malhotra, "Response of Salt Marshes to Wave Energy Provide Guidance for Successful Living Shoreline Implementation," *Living Shorelines: The Science and Management of Nature-Based Coastal Protection,* edited by Donna Marie Bilkovic, Molly M. Mitchell, Megan K. La Peyre, and Jason David Toft (CRC Press, 2017), 214.
3. Melinda K. Ehrich, and Lora A. Harris, "A Review of Existing Eastern Oyster Filtration Rate Models," *Ecological Modelling* 297 (2015): 201–12, https://doi.org/10.1016/j.ecolmodel.2014.11.023.
4. Jamie G. Boswell, Judy A. Ott, Anne Birch, and Daniel Cobb, *Charlotte Harbor National Estuary Program Oyster Habitat Restoration Plan* (Charlotte Harbor National Estuary Program Technical Report, December 2012), 8. http://chnep.wateratlas.usf.edu/upload/documents/134_CHNEP-Oyster-Restoration-Plan-12Dec2012-lowres.pdf.
5. Ido Sella, Tomer Hadary, Andrew J. Rella, Bernhard Riegl, Denise Swack, and Shimrit Perkol-Finkel, "Design, Production, and Validation of the Biological and Structural Performance of an Ecologically Engineered Concrete Block Mattress: A Nature-Inclusive Design for Shoreline and Offshore Construction," *Integrated Environmental Assessment and Management* 18, no. 1 (2022): 148–62, https://setac.onlinelibrary.wiley.com/doi/epdf/10.1002/ieam.4523; Reginald Davey, "Eco-Friendly Concrete Solutions for the Protection of Marine and Coastal Infrastructure," AZO Cleantech, April 22, 2021, https://www.azocleantech.com/article.aspx?ArticleID=1226.
6. C. Meyer, "The Greening of the Concrete Industry," *Cement and Concrete Composites* 31, no. 8 (2009): 601–5.
7. F. Winnefeld, "Calcium Sulfoaluminate Cement: An Example of a Low CO2-Alternative to Portland Cement," Paper presented at the WTA colloquium Effect of Climate Change on Built Heritage, Eindhoven, The Netherlands. March 11–12, 2010.
8. Sella et al., "Design, Production, and Validation."; Davey, "Eco-Friendly Concrete Solutions."
9. Ofer Berman, Michael Weizman, Asa Oren, et al., "Design and Application of a Novel 3D Printing Method for Bio-Inspired Artificial Reefs," *Ecological Engineering* 188 (2023): 106892.

10. University of Central Florida, "Coastal and Estuarine Ecology Lab (CEELAB)," accessed February 3, 2025, https://sciences.ucf.edu/biology/ceelab/; BESE-Products.com, "The Oyster Mat," accessed February 3, 2025, https://www.bese-products.com/article/biodegradable-oyster-mats/.
11. Linda J. Walters, Annie Roddenberry, Chelsey Crandall, et al., "The Use of Non-Plastic Materials for Oyster Reef and Shoreline Restoration: Understanding What Is Needed and Where the Field Is Headed," *Sustainability* 14, no. 13 (2022): 8055, https://doi.org/10.2290/su14138055.
12. Walters et al., "Use of Non-Plastic Materials for Oyster Reef."
13. Christine Angelini Ada Bersoza, Deidre Herbert, Emily Astrom, et al., *Living Shorelines: A Manual for Re-Engineering Living Shorelines to Halt Erosion and Restore Coastal Habitat in High-Energy Environments,* Technical Report (National Estuarine Research Reserve System Science Collaborative, University of Michigan, 2018), https://graham.umich.edu/media/files/Project_Manual_Living_shorelines_high_energy_enviro.PDF.
14. Brevard Zoo, "Seeing Early Success in Plastic-Free Oyster Restoration Material," *Behind the Scenes* (blog), accessed February 3, 2025, https://brevardzoo.org/news/early-success-plastic-free-oyster-restoration/.
15. Mohammad J. Uddin, Kelly J. Smith, and Craig W. Hargis, "Development of Pervious Oyster Shell Habitat (POSH) Concrete for Reef Restoration and Living Shorelines," *Construction and Building Materials* 295 (2021): 123685, https://www.sciencedirect.com/science/article/pii/S0950061821014458.
16. Hunter Mathews, Mohammad J. Uddin, Craig W. Hargis, and Kelly J. Smith, "First-Year Performance of the Pervious Oyster Shell Habitat (POSH) Along Two Energetic Shorelines in Northeast Florida," *Sustainability* 15, no. 9 (2023): 7028, https://doi.org/10.3390/su15097028.
17. Uddin et al., "Development of Pervious Oyster Shell Habitat"; Hunter Mathews, Lauren Cope, Jacob Waggoner, et al. *The Pervious Oyster Shell Habitat (POSH): Introduction and Construction Manual* (University of North Florida, 2024), http://dx.doi.org/10.13140/RG.2.2.20903.56489. The construction manual can be downloaded for free.
18. Becca Hatchell, Katie Konchar, Maria Merrill, Colin Shea, and Kent Smith, "Use of Biodegradable Coir for Subtidal Oyster Habitat Restoration: Testing Two Reef Designs in Northwest Florida," *Estuaries and Coasts* 45, no. 8 (2022): 2675–89, https://link.springer.com/article/10.1007/s12237-022-01094-6.
19. "Oyster Prisms and Volcanoes to Help the Lagoon," *Florida Today,* July 2, 2022, https://www.floridatoday.com/picture-gallery/news/2022/07/02/concrete-oyster-prisms-lagoon/7795833001/.
20. Walters et al., "Use of Non-Plastic Materials for Oyster Reef."
21. C. C. Atkinson, "Quantifying Oyster Reef Development and Nekton Use of Habitat Restored Using Traditional and Alternative Substrates in a Georgia (USA) Estuary," (PhD diss., Savannah State University, 2022); Emory H. Wellman, Christopher J. Baillie, Brandon J. Puckett, Sarah E. Donaher, Stacy N. Trackenberg, and Rachel K. Gittman,

"Reef Design and Site Hydrodynamics Mediate Oyster Restoration and Marsh Stabilization Outcomes," *Ecological Applications* 32, no. 2 (2022): e2506, https://doi.org/10.1002/eap.2506.

22. Chelsea K. Nitsch, Linda J. Walters, Joshua S. Sacks, Paul E. Sacks, Linda G. Chambers, "Biodegradable Material for Oyster Reef Restoration: First-Year Performance and Biogeochemical Considerations in a Coastal Lagoon," *Sustainability* 13, no. 13 (2021): 7415, https://doi.org/10.3390/su13137415.
23. Devin Comba, Terence A. Palmer, Natasha J. Breaux, and Jennifer B. Pollack, "Evaluating Biodegradable Alternatives to Plastic Mesh for Small-Scale Oyster Reef Restoration," *Restoration Ecology* 31, no. 3 (2022): e13762, https://doi.org/10.1111/rec.13762.
24. Nature Conservancy, "Mangroves Reduce Flood Damages During U.S. Hurricanes, Saving Billions of Dollars in Property Losses," accessed October 29, 2019, https://www.nature.org/en-us/newsroom/mangroves-reduce-florida-flood-damages/.
25. Caitlin M. Snyder, Laura C. Feher, Michael J. Osland, Christopher J. Miller, A. Randall Hughes, and Karen L. Cummins, "The Distribution and Structure of Mangroves (*Avicennia germinans* and *Rhizophora mangle*) Near a Rapidly Changing Range Limit in the Northeastern Gulf of Mexico," *Estuaries and Coasts,* 45, no. 1 (2021): 181–95, https://doi.org/10.1007/s12237-021-00951-0.
26. Paul L. Knutson, Robert A. Brochu, William N. Seelig, and Margaret Inskeep, "Wave Damping in *Spartina alterniflora* Marshes," *Wetlands* 2 (1982): 87–104, https://doi.org/10.1007/BF03160548.

# 6

# Coastal Living Shorelines

## *Planning a Shoreline*

**PLANNING A LIVING SHORELINE** involves several key stages that help guide the process of designing and implementing this environmentally friendly approach to coastal protection. This chapter covers the crucial aspects of planning, from understanding the site to working with restoration professionals and obtaining necessary permits. It discusses how to make decisions about what can be constructed, how to set goals, and when to call for professional help. It also outlines the individual steps of the planning process and whether considering a do-it-yourself (DIY) approach or collaborating with experts is appropriate. The planning stages outlined below will serve as a comprehensive guide to set the stage for success of the living shoreline project. With proper planning, building a living shoreline will be a smoother process.

### Choosing the Right Living Shoreline for the Property

The best type of living shoreline will depend on several factors, including the size of the property, the type of shoreline, and the amount of erosion, both past and ongoing. The aesthetic value of the different types of living shorelines is also an important consideration.

If homeowners or other shoreline managers are unsure about which type of living shoreline is right for their specific property, it is a good idea to hire professionals or experts who can help guide them through the process. A professional can help assess the property and recommend the best type of living shoreline for that property. This chapter describes the role of the experts in living shoreline projects and the process involved in the planning, designing, and permitting.

## DIY: Is It Possible for Homeowners to Do the Work Themselves?

Although it is possible to create a DIY living shoreline, many homeowners or other shoreline managers choose to work with a team of restoration professionals to streamline the process. The most difficult parts of the process are understanding the site and what can be accomplished there as well as getting through the permitting process. Some of the guesswork can be reduced if there is a living shoreline at a neighboring property that can serve as an example. Even if homeowners choose to do some of the work themselves, they may find it more efficient to put together a team who can assist them through the process, from planning and permitting through installation.

## Learn About the Site and Set Attainable Goals

The first step is to learn as much as possible about the property and the shoreline to protect, including the following:

- The size and slope of the property
- The type of soil
- The amount of wave energy
- The presence of any existing structures
- The types of plants and animals which are present on the shoreline and which of them are native, nonnative, or invasive (See chapter 4.)

Before starting, make sure to collect important information about the property (location, tax parcel ID, and GPS coordinates). Homeowners and property managers should keep a copy of the land deed to the property and a current survey on hand. If the property line extends to the mean high water line (MHW), then in most cases the owner is allowed to access and manage, within legal limits, that shoreline as part of his or her riparian rights. Tools such as Google Earth provide an aerial view of the property and adjacent areas; these views often provide key context for living shoreline planning. Equipped with this information, the homeowner will have a clearer understanding of what is allowable for their specific shoreline.

Note any structures or materials on the shoreline including seawalls, bulkheads, riprap, or other materials that may need to be removed or modified. Also note other municipal structures like stormwater outfalls and determine if there are municipal rights-of-way associated with those structures. Existing structures can make a difference in whether the living shoreline installation uses a nature-based or hybrid system and will influence the living shoreline design. If the property is on a canal, there may be limitations because of the often-narrow lateral space available in the canal. If there is enough width, riprap might be added at the foot of the seawall, which could be constructed to include planters for mangroves or saltmarsh grasses.

Defining the goals for a living shoreline project is another vital step. (See chapter 4, on setting goals.) Once there is a set of goals and a list of expectations for what the living shoreline is to accomplish, it can also be helpful to make a sketch of the shoreline with its projected living components. If additional help is needed in defining project goals, call in a local expert to help. Local experts may be found through the local extension office or environmental nonprofit groups.

## Working with Restoration Professionals: What Are Their Roles?

Whether homeowners or property managers plan on spearheading the project on their own or putting together a team to do the work, the first step is to call in a local expert who is familiar with the location. A local expert can fast-track the homeowners' or property managers' knowledge about the living shoreline process.

The local expert can also help guide and educate any contractors who may work on the project, especially if living shorelines are new to them. They can communicate and help translate the homeowners' or property managers' vision and goals to contractors and permitting agencies, so that goals are implemented as originally envisioned. A local expert will visit the site and guide homeowners through site-specific options for their shoreline. They may suggest additional restoration professionals who may also become an important part of the team. Collaboration with restoration professionals is often advantageous, as they possess specialized knowledge and experience that can streamline the process. The team of professionals usually includes an ecologist, a biologist, or an environmental consultant. For a directory of local experts in Florida visit the Florida Living Shorelines website at https://floridalivingshorelines.com/contacts/.

A team of restoration professionals will help with the following:

- Assessing the shoreline site
- Obtaining the necessary permits
- Choosing the right plants and materials for the shoreline
- Installing the living shoreline components correctly

These professionals perform tasks like creating a site assessment or completing a seagrass survey. They also might be involved in planning which plants, reefs, or submerged aquatic vegetation (SAV) are appropriate for the shoreline.

## Marine Contractors

Marine contractors are responsible for the site work and installation of the project unless the project is small and can be done by homeowners or volunteers. Typically, marine contractors will also be part of the planning process to develop a project timeline, estimate costs, obtain needed permits, and find local sources for materials like riprap. Some marine contractors have site engineers and ecologists on staff. Ideally, it is best to work with a contractor who is familiar with living shorelines.

So they can play a more effective role in improving Florida's shoreline properties, many contractors have taken the Florida Living Shorelines Training for Marine Contractors Course (https://www.flseagrant.org/workforce-training/living-shorelines-training-for-marine-contractors/) or the Florida Master Naturalist Program course in Coastal Shoreline Restoration (https://masternaturalist.ifas.ufl.edu/media/masternaturalistifasufledu/CSR_Course-Description.pdf).

For a list of contractors who have taken the Living Shorelines for Marine Contractor Training course, visit https://floridalivingshorelines.com/businesses/.

## Landscape Architects and Other Landscape Professionals

Homeowners and other shoreline managers can also consult a landscape architect if they are interested in changing the area of the property that is above the MHW line. Landscape architects plan how a space is used, how it functions, and how it looks. They can integrate the rest of the yard with a living shoreline, especially when building upland buffers, by adding appropriate native plants that will help keep stormwater and pollutants like fertilizer from running into the water from the upland landscapes. As with marine contractors, ideally, find a landscape architect familiar with native plants, upland buffers, living shorelines, and their overall goals. To find members of the Florida Association of Native Nurseries, go to the Plant Real Florida website at https://www.plantrealflorida.org. (See chapters 8 and 9 for more details on upland buffers and landscaping above the shoreline.)

## LIVING SHORELINE TRAINING AND EDUCATION

### LIVING SHORELINE TRAINING FOR MARINE CONTRACTORS

The Florida Sea Grant program and the University of Florida Institute of Food and Agricultural Sciences (UF/IFAS) teaches a two-day living shoreline course directed toward professionals seeking to add living shoreline practice to their businesses. The target audiences for the course are environmental consulting firms, marine construction businesses, landscape and nursery professionals, and restoration practitioners from local governments, colleges and universities, and nongovernmental organizations (NGOs). Additionally, the course has been attended by local, state, and federal government agencies that may be permitting or managing living shoreline projects.

The course is also open to anyone from the general public who is interested in living shorelines. Topics include site assessment for living shorelines; project design; state and federal permitting for living shorelines; costs, materials, and techniques for living shoreline construction; project evaluation; and maintenance. Participants are taught through classroom lectures and exercises, field trips, and self-directed written and field assignments. Course graduates are eligible to receive continuing education units (CEUs) for their professional practice.

For more information on the course schedule and locations, visit Florida Living Shorelines at https://www.floridalivingshorelines.com.

### FLORIDA MASTER NATURALIST PROGRAM— COASTAL SHORELINE RESTORATION

UF/IFAS also offers a special topics course, Coastal Shoreline Restoration (CSR), within their popular Florida Master Naturalist Program (FMNP). This adult education course is targeted more toward the public, but professionals and other restoration practitioners have found it useful to attend this course as well.

> The mission of the Florida Master Naturalist Program (FMNP) is to promote awareness, understanding, and respect of Florida's natural world among Florida's citizens and visitors.

The FMNP CSR course consists of 24 instructional hours experienced through lectures, field trips, and other hands-on activities designed to help participants learn about natural estuarine shoreline habitats and nature-based alternatives to seawalls and other gray infrastructure for shoreline protection and restoration. Topics covered include oyster, mangrove, salt marsh restoration, and the integration of those restored habitats through living shorelines. Past graduates have used the course information to initiate restoration projects as professionals or volunteers in their communities and for environmental education and interpretation.

Course graduates receive a Certification of Course Completion and the FMNP embroidered patch and course-specific lapel pin.

For more information on this course, other FMNP courses, and the current schedule of offerings please visit the Florida Master Naturalist Program page at https://masternaturalist.ifas.ufl.edu/.

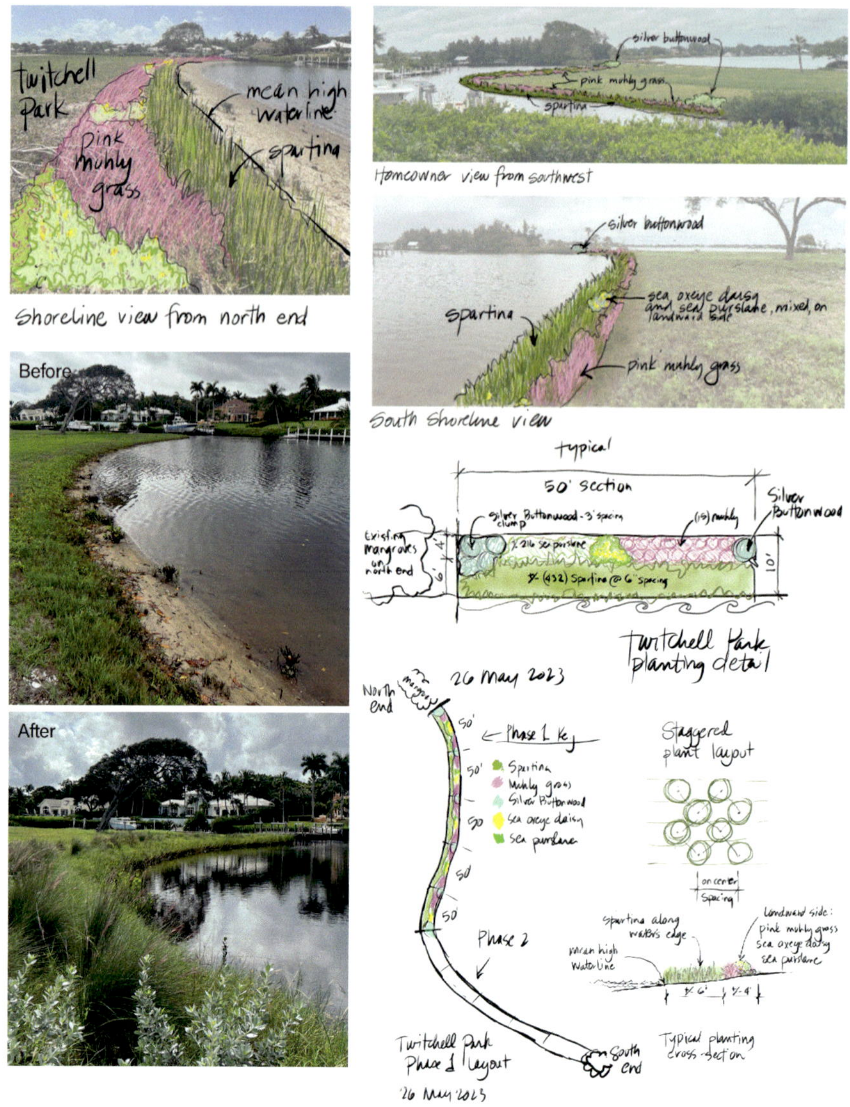


**FIGURE 6.1** Landscape design sketches and site photos from a completed living shoreline project in Martin County, Florida, by landscape architect Meg Whitmer. *Sketches and photos: Ocean Research and Conservation Association Inc. (ORCA).*

Landscape companies may also be among the professionals installing the living shoreline components. The primary contractor for the living shoreline may subcontract a company that would be responsible for sourcing and installing native plants from the homeowner's or shoreline manager's plant list. A commercial source of plants from licensed nurseries is generally required to obtain a permit. (See the permitting section below.) Another important aspect related to working with landscapers is that they may be the main professionals maintaining the vegetation on the living shoreline and in the upland buffers. Maintenance would include ensuring installed plants are successfully established with initial watering, replacing those that are not successful, and controlling and removing invasive plant species (also required by permit) and other weeds that might out-compete the newly planted living shoreline plants. Landscapers, probably more than the other professionals described in this chapter, understand and regularly perform maintenance as part of their daily job activities. Once the living shoreline is installed, they can work with the homeowners or property managers to establish a maintenance contract to ensure the long-term success of the project. Recent efforts to increase living shoreline education among professionals have targeted the landscape industry to increase the population of landscapers familiar with living shorelines.

## Local Conservation Groups

In some cases, homeowners and property managers may be able to partner with a local conservation group. These groups can help with locating materials like oyster shells, and they can coordinate volunteers to help install oyster reefs or plants. Several coast-based conservation groups have experts and experience with living shorelines and habitat restoration. (See chapter 2.) An individual from one of those groups may also be the local expert providing that initial guidance, bringing together other professionals to work on the project, or even being involved in the installation of the living shoreline.

## County Extension Offices

The University of Florida Institute of Food and Agricultural Sciences (UF/IFAS) extension service and Florida Sea Grant have been instrumental in bringing living shoreline education to local communities and professionals through their respective County Extension Offices. (See the sidebar titled "Living Shoreline Training and Education.") The local Florida Sea Grant extension agent can help homeowners find local experts. Florida Friendly Landscaping and Horticulture extension agents can assist in finding sources of native plants and landscape professionals. Also, Master Gardener groups and Florida Native Plant Society chapters can provide useful information on coastal plants that can be part of a living shoreline "seascape." If the local county extension office does not have specific expertise on living shorelines, they can help homeowners find the nearest one that does.

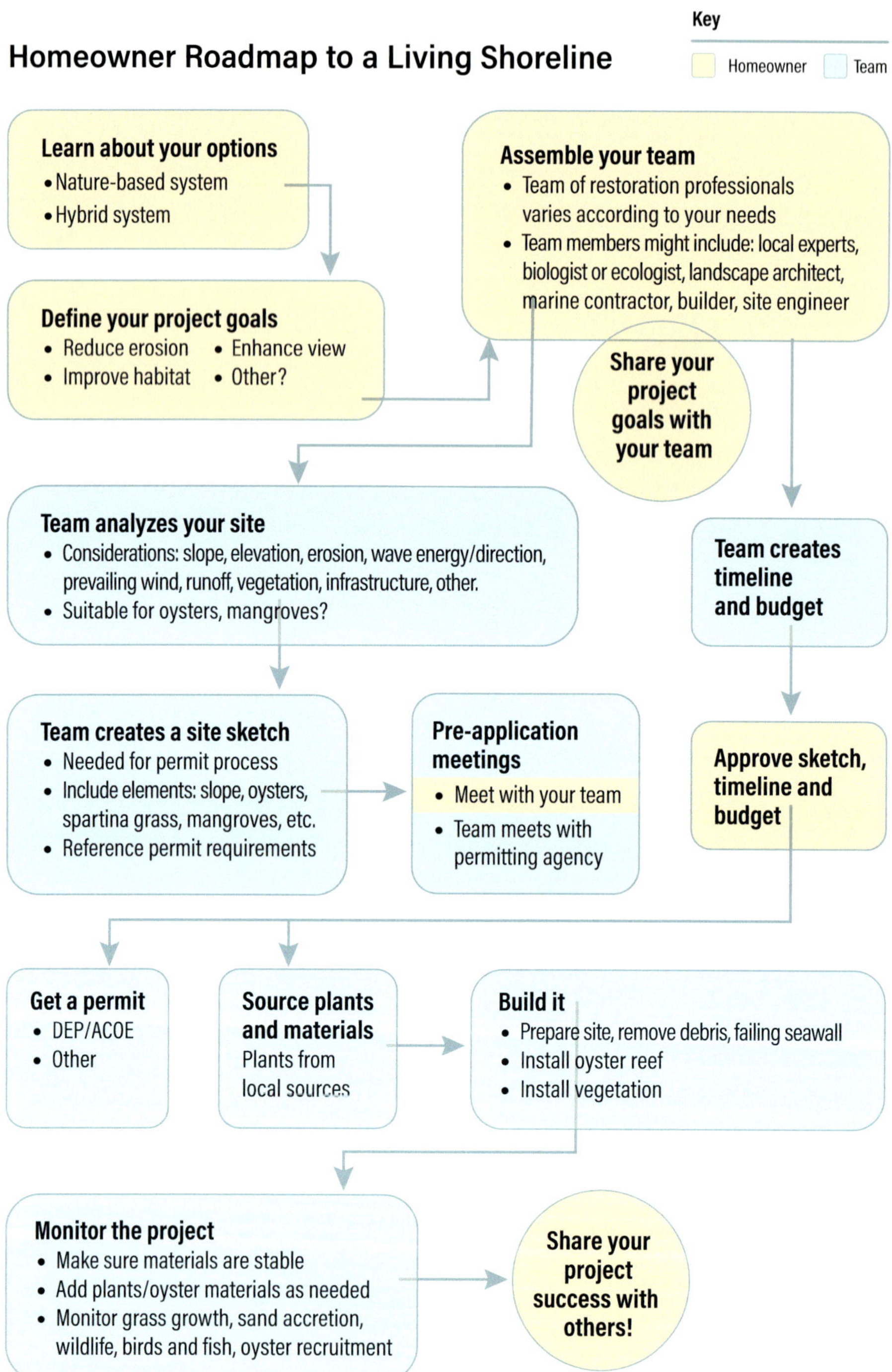


**FIGURE 6.2** The steps involved in planning and building a living shoreline. *Illustration by Marjorie Shropshire.*

## The Process: From Site Assessment and Project Permits to Sourcing Materials

Once the team is in place, the team members will follow the steps detailed below to get the shoreline project started. Homeowners or other shoreline managers will need to communicate with the team members to approve each of the steps. Homeowners, property owners, and the beneficiaries of a living shoreline are encouraged to be involved as much as possible (depending on their abilities) in the process so that their goals are met, their concerns are answered, and their team of experts explains their rationale behind designs and any potential changes to the original plans. This involvement is also educational, so the property owners or managers can become better advocates for living shorelines.

### Step 1. The Initial Planning Meeting(s)

Once the team of restoration professionals has been assembled, the homeowners or shoreline managers should arrange a first meeting, preferably on-site, to discuss their vision and goals for the project and any concerns (e.g., extreme erosion or flooding, threats to existing structures, recreational access, etc.). This initial meeting is important to ensure the experts agree with each other and with the homeowner or property manager. If there are disagreements on how to approach the living shoreline project, whether technical, regulatory, or aesthetic in nature, this is the time to address them and come to a consensus. Homeowners or shoreline managers should plan on having more than one meeting as the project planning continues, to receive and share updates and any potential changes among all members of the team.

### Step 2. Assessing the Site

Team members will visit the shoreline and assess it for factors like erosion, slope, wave energy, boat wakes, water depth, underwater profile, and the presence of oysters and seagrass. They will also look at what is on neighboring shorelines and will take into consideration any hard structures like riprap, bulkheads, seawalls, or docks.

### Step 3. Site Sketches

A set of six sketches will be needed for the project:

- A vicinity map with enough detail to allow someone to find the project
- An overlay map showing the delineation of all wetlands and surface waters on-site
- An overhead (plan) view of the *present* site condition
- Cross-section drawings of the *present* site condition
- An overhead (plan) view of the *proposed* project
- Cross-section drawings of the *proposed* project

See examples of site sketches in figure 6.4 for corresponding example images.

These sketches are required by the permitting agencies, and they are also helpful for planning and getting on the same page about the project vision with the team of contractors. Some of the information collected during the site assessment will be included in the sketches. The plan view and cross-section sketches will also specify the size and placement of natural elements the team recommends, like plants and oyster reefs. Make sure that all items in these sketches are clearly labeled and that appropriate disclaimers such as "drawing not to scale" are included. The team will also create a written description of the site, what they plan to do, and how the project will be implemented, including what tools and materials will be used, the project timeline, the dimensions of the project elements, and the justification for the proposed approach.

This written narrative will also be needed for the permit process and is a supplement to the site sketches. See below for a list of information that should be included on the site sketches; and be sure to consult with permitting guidelines to ensure that all necessary information is included. Failure to include all the necessary information will result in delays in permitting.

### THE VICINITY MAP

- Overhead map view of the property showing the entire parcel, neighboring parcels, and the nearest access street
- Include enough detail to allow someone to find the project site
- Indicate the location of the property (e.g., a star or box around the parcel where the project will be installed)

Consider using free tools such as Google Maps or Google Earth to produce the vicinity map.

### THE OVERLAY MAP

The overlay map can be produced using publicly available data from the US Fish and Wildlife Service (USFWS). Access the USFWS National Wetlands Inventory Wetland Mapper Tool (https://www.fws.gov/program/national-wetlands-inventory/wetlands-mapper) and zoom in to the specific site location. Export the map using the print functions of the tool or by taking a screenshot. Include enough information to verify the location of seagrasses or wetlands. The best practice is to match the map extent to the vicinity map.

### DETAILED SITE SKETCHES (PLAN AND CROSS-SECTION VIEWS)

The detailed site sketches should depict both the current site conditions in cross-section and in overhead (plan) view (first set) and the proposed project in cross-section and in overhead (plan) view (second set). Multiple cross-section sketches may be needed for larger projects, but most small projects only require one.

Each of these drawings must show the following:

- The approximate location of the mean high water (MHW) and mean low water (MLW) lines
- The linear distance along the shoreline covered by the project, labeled in feet
- The linear distance the project will extend into the water, as measured from the mean high water (MHW) line, labeled in feet
- Planned spots for placement of plants, breakwater/sill elements, and gaps in breakwaters with dimensions such as linear distances and/or heights noted, labeled in feet
- Existing structures (e.g., docks, sheds), labeled as such
- Any nearby habitats
- The approximate location of the shoreline/property boundary of neighboring properties
- An arrow indicating the direction of north
- The location(s) of the cross-section drawing(s) shown on the plan view drawing, usually denoted by drawing a dashed line across the area and labeling it as "cross-section A," or "cross-section B." If creating multiple cross-sections, be sure to label each cross-section drawing to match the names in the plan view

The homeowner or shoreline manager may produce these sketches either by hand or by using simple shape illustrations available in computer programs such as Microsoft PowerPoint. Drawings need not be to scale as long as this is noted on the drawing.

## Shoreline Site Assessment

### SITE CONDITIONS

Notes from Desktop Analysis:

Navigational channel more than 300 feet away from shoreline, soil type sandy, several water quality monitoring stations nearby (area has high nitrogen concentrations). Historically, oyster reefs thrived in the area. Prevailing wind direction is from the SW and intermittent high winds (20+ mph) occur at least monthly. Property is 150 linear feet.

Erosion: ☐ None ☐ Light ☒ Moderate ☐ Severe

Source of erosion (Upland runoff, storm events or major disturbances, undercutting, wave/wind induced)

Wind waves primarily, some boat traffic on weekends and larger boats in channel

Existing shoreline structures and shoreline type (e.g. seawall, bulkhead, rubble, vegetation, pipes, dock, kayak access, etc.)

Eroding vegetation, dock on south end of property

Wind exposure (approximate distance and direction that shoreline faces)

Faces SW, 2.5 mile fetch

Prevailing winds at time of assessment Southwesterly, 15 mph

Energy intensity: ☐ Low ☒ Medium ☐ High

Nearshore (1 m off shoreline) water depth 24-36 inches (moderate)

Tidal fluctuation 3 feet Tidal at time of assessment Low

Boating activity

- ☐ None or paddlecraft only
- ☐ Minimal (occasional small, motorized watercraft)
- ☒ Moderate (regular traffic, small and large motorized watercraft)
- ☐ Heavy (regular traffic including yachts, commercial vessels, tugs, barges)

Proximity to navigational channel 350+ feet

Safety considerations None

**FIGURE 6.3** Example of basic site assessment details to be collected. These observations will be used by the project team for planning the appropriate living shoreline configuration and building blocks. *From https://www.flseagrant.org/workforce-training/living-shorelines-training-for-marine-contractors/*

### UPLAND FACTORS

Existing bank height and slope

1.5 - foot scarp, bank sloped at 7:1

Sources of freshwater runoff/outfall

Rain runoff from uplands (turf grass lawn)

Sediment (e.g. sandy, peat, clay, organics, etc.)

Sandy with some organics in section of high marsh vegetation that hasn't yet eroded

### ECOLOGICAL FACTORS

Oysters present? (Yes)/ No  Attached to: Dock pilings

Submerged aquatic vegetation? Yes /(No)

Shorebird useage? ☐ Feeding ☐ Nesting ☐ Loafing ☐ Other

None observed

Existing vegetation – native and exotic, upland and wetland:

Some remnant high marsh (*Spartina patens*) with exposed roots and signs of erosion. No exotic vegetation noted. Upland vegetation is manicured turf grass

Water quality/salinity

Elevated nitrogen in the area, has suffered from algal blooms and red tide intermittently. Salinity 14-24 ppt

Nearby sources of freshwater None

Approximate wetland acres to be affected

To be determined based on property owner consultation about options and permit pre-application meeting

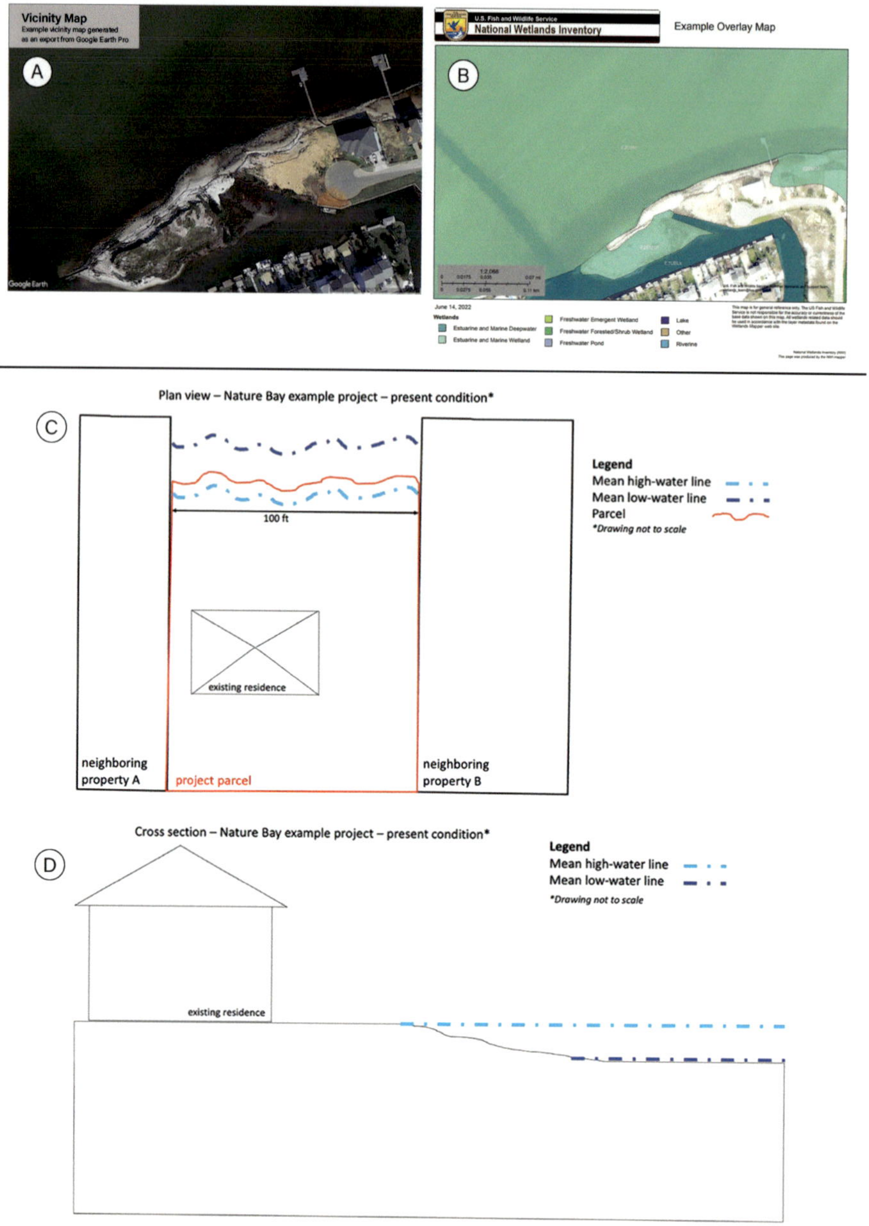


**FIGURE 6.4** Examples of site sketches required for permitting including present condition. *(A)* A vicinity map with enough detail to allow someone to find the project; *(B)* An overlay map showing the delineation of all wetlands and surface waters on site; *(C)* An overhead (plan) view of the present site condition; *(D)* Cross-section drawings of the present site condition; *(E)* An overhead (plan) view of the proposed project; and *(F)* Cross-section drawings of the proposed project. *Reproduced with permission from Savanna Barry, Sara Martin, and Eric Sparks, "Homeowner's Guide to the Living Shoreline Permit Exemption Part 2: United States Army Corps of Engineers," EDIS Publication SG 189, UF/IFAS, June 2022 (original publication March 2019), https://doi.org/10.32473/edis-sg189-20192023.*

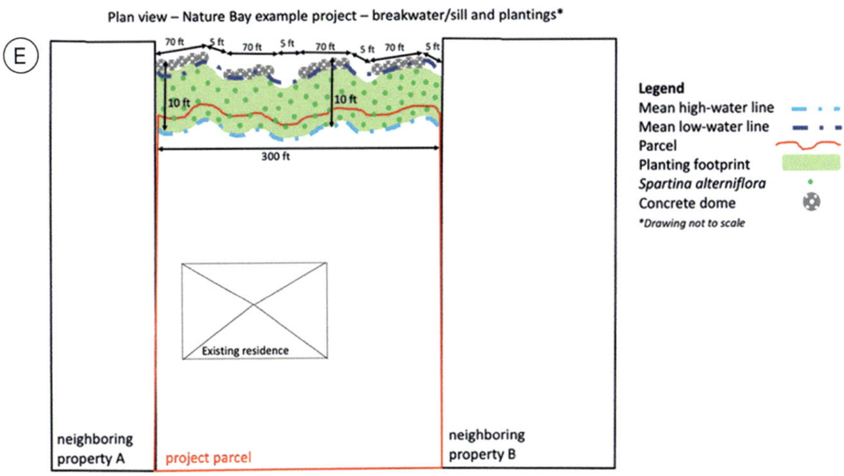


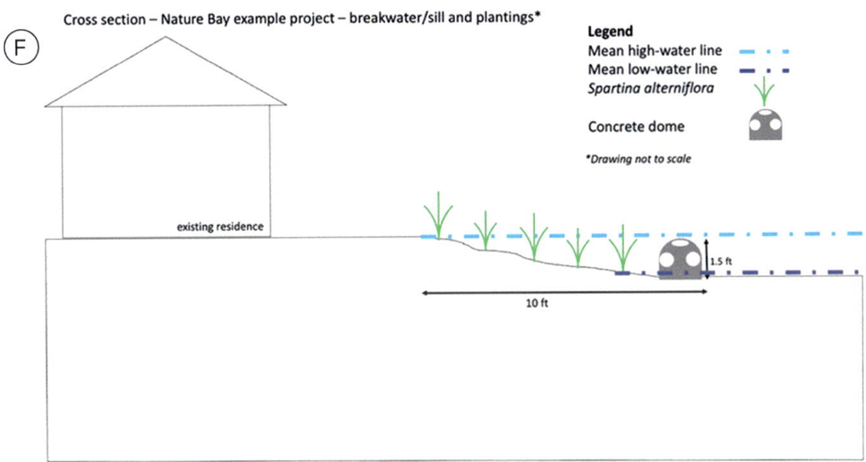


FIGURE 6.4 CONTINUED

It may be helpful to include extra photographs of the site to aid in the explanation. Or a contractor could provide the drawings as part of the contract, but this is typically not necessary for simple projects. If the budget allows, then an engineer can produce site (current and proposed) drawings to exact measurements. This may be especially important for hybrid living shorelines, which include engineered structures such as seawalls, bulkheads, or riprap. This would be necessary if those structures were planned as new constructions. In many hybrid living shoreline retrofit projects, natural elements are incorporated into existing gray infrastructure.

In such cases, engineer drawings are likely already available from when those structures were first planned and permitted.

## Step 4. Team and Permit Pre-Application Meetings

Next, the team will plan an internal meeting before meeting with state and federal agencies. During this meeting, the team will share the plan that they will bring to the agencies. Now is the time for the homeowner to ask any questions about what will be created on their shoreline and what will be requested during the permitting process. The team will then have a pre-application meeting with permitting agencies to make sure everything is in order and to help determine what permits may be needed. This pre-application meeting will be primarily with the Florida Department of Environmental Protection (FDEP), but hopefully US Army Corps of Engineers (USACE) staff can be present as well, or a meeting can be arranged with them soon after. The pre-application meeting will, in most cases, be between the lead contractor and FDEP, but including key subcontractors can be helpful.

Homeowners or shoreline managers may not need to attend this meeting, but it is highly recommended that they do so, because it will further clarify the process for them, and they will know firsthand if the government agencies have any concerns about their project. The team should be able to effectively convey and explain agency comments, especially those that may be unclear.

The pre-application meeting is extremely important because it gives the team the opportunity to address any questions that the FDEP and USACE might have prior to submitting a permit application. This meeting should be held as early as possible in the planning process, before hard decisions are made about the design of the living shoreline. This will ensure that the permit application is as complete as possible and avoid potential extensive Requests for Additional Information (RAIs) (see the section on permitting below) when the FDEP sends the application out for other government agencies' comments. To start the pre-application and permitting process, contact your FDEP district office. (See the link at the end of this chapter.)

## Step 5. Create a Timeline and a Budget

The team will also create a timeline and budget, so homeowners are aware of the schedule and costs. The length of time that will be required upfront is surprisingly long, and some permits take longer than others. For instance, if a seagrass survey is required, it can be performed only during the seagrass growing season (April through September) and, therefore, it can hold up the process if the team must wait for the survey window to open. Discuss the plan with the team. Costs will vary according to the specifics of the project, but living shorelines are significantly less expensive than seawalls.

## Step 6. Approve the Plan

The team should update site sketches and the project narrative to reflect any changes that were agreed upon during the pre-application meetings. Once the plan, timeline, and budget are completed, the homeowners or shoreline managers will need to approve the team's plan. When everyone is on the same page, the team will apply for permits.

## Step 7. Project Permits and Regulations

Many homeowners and property managers are concerned with permitting, but this is where a team of experts is important. Hired professionals will generally take care of the entire permitting process. Some permits can go through on a fast track, while others might require more time.

In Florida, homeowners will need to obtain a permit from the FDEP before installing a living shoreline. The FDEP will review the plans and make sure that the living shoreline meets all the necessary requirements. In some parts of the state, FDEP has delegated the authority related to shoreline projects to the regional water management district (WMD). If this is the case for the area of the team's project, their permit application would be reviewed by the WMD staff on behalf of FDEP.

Depending on the location of the property, there may also be federal regulations that come into play. For example, if the project is located within navigable waterways, a permit from the USACE may be required. In some cases, FDEP can issue a bundled permit that also includes permission from the USACE. In other cases, the team may have to apply separately to both agencies.

For small projects where the shoreline is less than 500 feet (152.4 m) wide and extends no more than 10 feet (3.1 m) toward the water from the mean high tide line, an FDEP permit exemption may be available.

### LOCAL RULES AND PERMITS

Be sure to consult local county and/or city governments in case a local permit is required. Living shoreline projects may fall under local regulations and codes concerning coastal development and shoreline stabilization. Staff from a local government's building code or growth management division may be the appropriate contacts to determine whether any local permits may be required. This is important for the team to be aware of, as local regulations may vary widely among counties and municipalities. If local governments are not consulted, a contractor familiar with regulations in one county may encounter obstacles or even restrictions to implementing projects in another county.

**TABLE 6.1** Thresholds for state and federal living shorelines permitting

| Project scale | Target thresholds | Federal permit instrument | State permit instrument |
|---|---|---|---|
| Small-scale | • No more than 500 linear feet (152.4 m) along shoreline<br>• Extends no more than 10 feet (3.0 m) offshore from the mean high-water line<br>• Top of reef structures not higher than mean high water<br>• No import of fill other than soil imported with plantings | Nationwide permit (likely NWP 54) | Living shoreline exemption |
| Moderate-scale | • No more than 500 linear feet (152.4 m) along shoreline<br>• Extends no more than 30 feet (9.1 m) offshore from the mean low water line<br>• Import of fill is the minimum necessary to support plantings | Nationwide permit (likely NWP 54) | Individual permit |
| Large-scale | • Exceeds any of the above thresholds | Standard individual permit | Individual permit |

*Note:* Regulatory thresholds may change and, in some cases, permitting officials may have the authority to waive certain limits. There are also other considerations and rules to follow other than project thresholds. Consult with the team and regulatory staff during a pre-application meeting to ensure the project meets targets for the most efficient permitting route.

**TABLE 6.2** Living shorelines: Relative comparisons

| Feature | Living shorelines | Armored shorelines |
|---|---|---|
| Initial cost | \$465–\$1,400* | \$1,200–\$2,775* |
| Maintenance cost | 10–20 percent of installation cost | 50–100 percent of installation cost |
| Longevity | Lasts for decades | Needs to be replaced every 20–30 years |
| Environmental impact | Less impact | More impact |
| Aesthetics | More natural | Less natural |

*Note*: *Average cost per square foot (0.1 $m^2$)

See the following links for more information about permitting:

A Homeowner's Guide to the Living Shoreline Permit Exemption, Part 1, https://edis.ifas.ufl.edu/publication/sg187

A Homeowner's Guide to the Living Shoreline Permit Exemption, Part 2, https://edis.ifas.ufl.edu/publication/sg189

Permitting Living Shorelines in Florida, https://floridadep.gov/rcp/florida-resilient-coastlines-program/documents/permitting-living-shorelines-florida

## Step 8. Sourcing Plants and Other Materials

While the permit process proceeds, the team will line up materials for the living shoreline project. They will locally source plants (smooth cordgrass and other marsh plants, red mangroves) and locate other materials for the living shoreline, such as oyster shells and coquina rock. (See chapter 5 for more details.)

## ▸ CASE STUDY ◂

### PROJECT #3. PRIVATE RESIDENCE AT SEWALL'S POINT LIVING SHORELINE

**Location:** Southeast Florida, Martin County, Sewall's Point, St. Lucie River

**Upland Property Ownership:** Private home

**Site History:** The shoreline experienced erosion from storms and boat wakes. A large escarpment (greater than 3 feet (0.9 m) high) resulted from historical and episodic erosion. Erosion extended into the upland turf area.

**Dates:** Phased oyster reef construction and plantings throughout 2016.

**Living Shoreline Type:** Natural

**Living Shoreline Building Blocks:** Native plants (upland buffer and some marsh plants); oyster reefs (bagged shell) placed offshore; large coquina boulders were used to fill in scarps from erosion to lessen slope.

**Size:** Length of shoreline is 100 ft (30.5 m); oyster reef area (fringing reef and patch reefs) is 1,130 sq. ft. (105 m); Upland planting area is 500 sq. ft. (46.5 m)

**Permitting:** Individual permits (both FDEP and USACE) were issued to Martin County as part of large-scale oyster restoration in the St. Lucie and Loxahatchee Rivers. As an additional component of the permits, smaller scale living shoreline sites were allowed with the approval of waterfront property owners, whose properties fronted these shoreline locations. A letter of support from each property owner modified the permit to allow the construction of oyster reefs on state-owned submerged land.

**Construction:** Plantings—high marsh vegetation, primarily salt meadow cordgrass (*Spartina patens*), was installed above the MHW line. Intertidal oyster reefs were constructed from bagged, recycled shells. Two types of reefs were constructed:

- a fringing breakwater reef to provide nearshore protection of the plantings, and
- four patch reefs farther offshore to provide additional shoreline protection and habitat.

**Maintenance:** Ongoing maintenance (initial irrigation, plant replacement) from the landscaping company was performed to ensure plant survival and success. Oyster reefs were inspected and monitored for spat settlement and survival.

**Cost:** Total cost—$20,000 over a period of seven years. $10,000 was devoted to the design, installation of plants, and the placement of natural rocks on the shoreline to stabilize the erosion. This also included maintenance of the plantings. $4,000 was devoted to the costs of installing the oyster reefs. After several years, an additional $6,000 was spent on adding more shoreline rocks and plantings.

**Notes:** This site was notable for its partnerships between a nonprofit conservation group with a local expert (the Florida Oceanographic Society [FOS]) who planned, designed, and installed oyster reefs, a local landscape architect (Meg Whitmer, who handled the landscape design and plant list), and a local landscaping company (Allgreens Inc., which installed coquina rocks and the plants and performed the maintenance). Also notable were the property owner's specific goals of protecting the shoreline and improving water quality and habitat for the river. The owner installed a wildlife camera to document wildlife use of the shoreline area.

**Challenges:** This site experienced freshwater discharges shortly after completion of the reefs in 2016 and later in 2018, which likely extirpated live oysters on the reefs. Coupled with these discharges were cyanobacterial blooms.

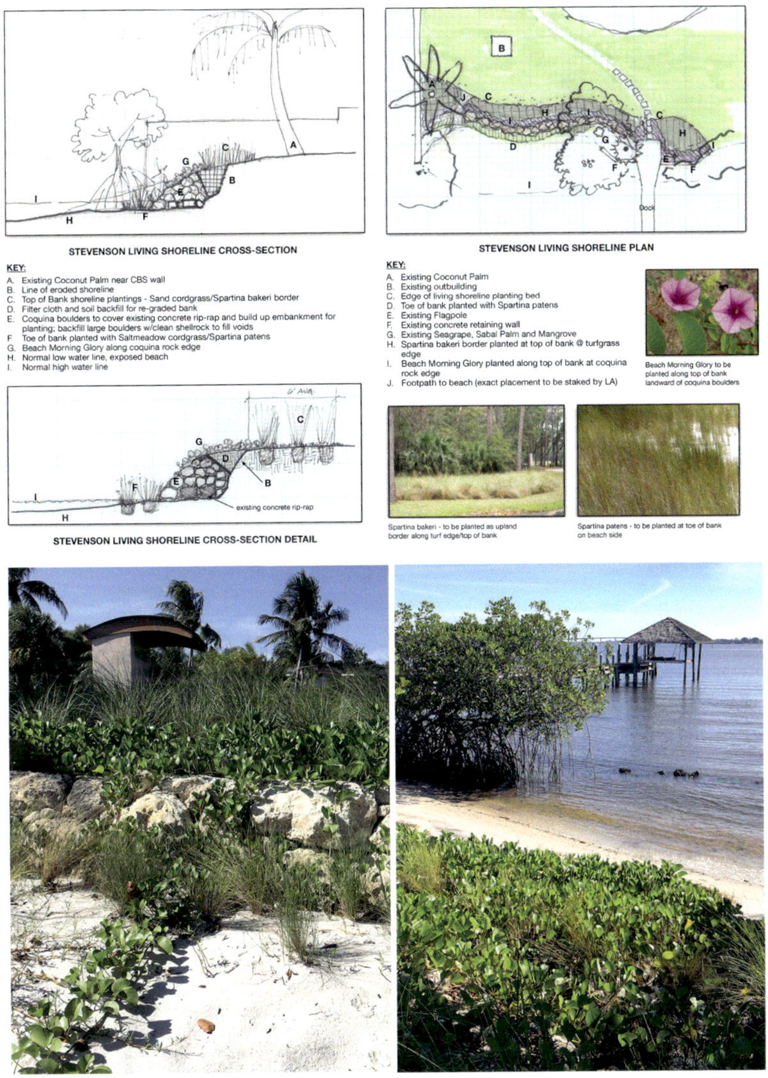


**FIGURE 6.5** Landscape design sketches and site photos from a completed living shoreline project in Martin County, Florida. *Cordgrass and beach morning glory photos by Shirley Ray Denton. Sketches and other photos by Meg Whitmer.*

## Relative Costs of Living Shoreline Options vs. Armored Shorelines

The cost of installing a living shoreline will vary depending on the size of the property, the type of living shoreline, and the cost of materials and labor. In general, living shorelines are more cost-effective than traditional seawalls. Recent estimates for Florida indicate living shorelines are roughly 8.5 times less expensive to install than armored shorelines.[1] Every shoreline is different because there are so many variables, but unlike seawalls, living shorelines improve and strengthen over time. Living shorelines and seawalls both need to be maintained, but the maintenance costs for living shorelines are significantly less expensive—about 2.7 times less expensive than seawall maintenance—based on a recent survey of Florida homeowners.[1] A seawall may need to be replaced every 20 to 30 years, but if designed and maintained properly, a living shoreline may never need to be replaced. Maintenance is generally simple and involves replacing plants, controlling invasive plants, and checking on the integrity of any hard sill or breakwater structures to determine whether any modules need to be added for replacement or to maintain their vertical position in the tidal zone.

## Conclusion

This chapter has covered the process of planning the living shoreline project, including assembling a team, assessing the site, determining the right type of living shoreline for the conditions, working with a team to develop a design, budget, and timeline, and submitting a permit application (and getting it approved!). The next step is to build the living shoreline.

## Notes

1. Savanna C. Barry, Laura K. Reynolds, Anna E. Braswell, Rachel K. Gittman, Steven B. Scyphers, and Ashley R. Smyth, "Perceived Effectiveness Drives Shoreline Decision-Making for Florida's Waterfront Property Owners," *Ocean and Coastal Management,* 258 (2024): 107353, https://doi.org/10.1016/j.ocecoaman.2024.10735

## Resources

Florida Department of Environmental Protection (FDEP), https://www.dep.state.fl.us/

Florida Department of Environmental Protection (FDEP), District Offices, https://floridadep.gov/districts#s

Florida (Jacksonville District) US Army Corps of Engineers offices, https://www.saj.usace.army.mil/Missions/Regulatory/Office-Locations/

Florida Living Shorelines, https://floridalivingshorelines.com/

U.S. Army Corps of Engineers, https://www.usace.army.mil/

# 7

# Coastal Living Shorelines

## *The Building Process*

**PREVIOUS CHAPTERS COVER THE** planning stages of creating a living shoreline, including setting goals, understanding the site, and assembling a restoration team. This chapter covers the exciting process of building a living shoreline, and it provides an overview of the steps for building two different living shoreline options. It will help property owners and shoreline managers understand the specific steps that need to be followed to construct a living shoreline.

### Building Process Overview

Once the building begins, things will usually progress quickly. Some initial site work will be done, and then the natural elements will be put in place, starting with an oyster reef or perhaps coquina rocks. Next, the living shoreline plants, which will be mostly marsh grasses and mangroves, will be added. Once the plants become established, they will fill in and, pretty soon, the homeowners will be able to sit back and enjoy the beauty of their new living shoreline.

### Building a Nature-Based Shoreline: From Site Prep to Installation and Monitoring

The following is an example of a typical nature-based living shoreline installation, but these steps might vary depending on the type of shoreline.

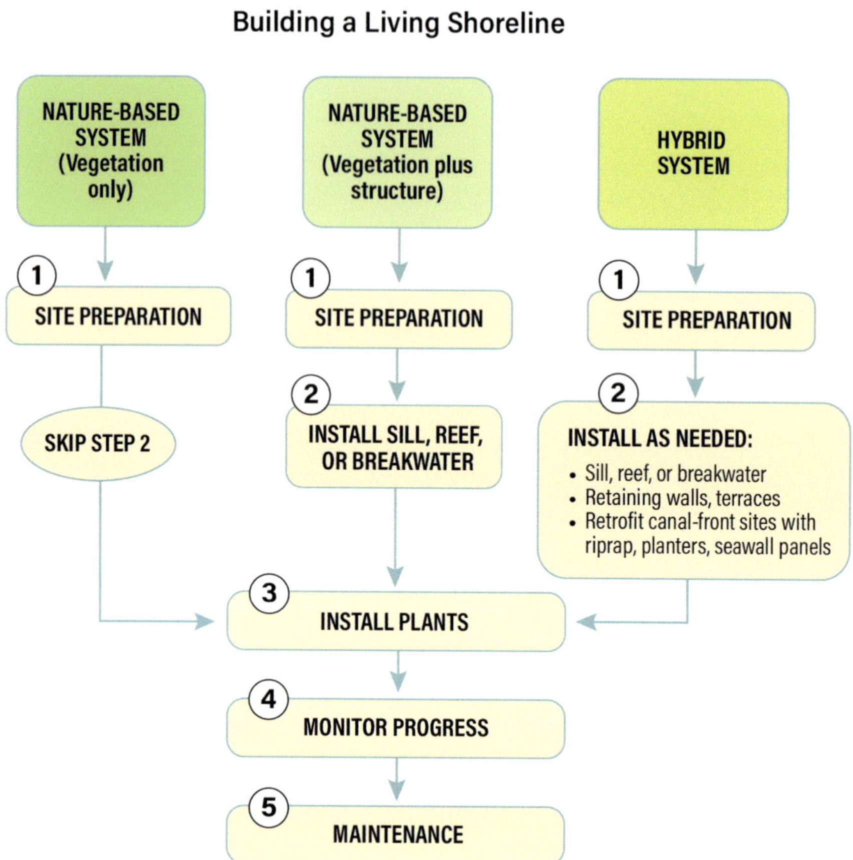


**FIGURE 7.1** The steps for building living shorelines with vegetation only, with vegetation and structure, and hybrid systems. *Illustration by Marjorie Shropshire.*

## Step 1. Site Preparation

Team members will already have assessed the shoreline. The first step involves any needed site work, such as removing invasive species or unwanted materials like concrete or other debris. Once the initial site work has been completed, construction of the living shoreline can begin.

If significant intervention is required to create a shoreline with a more gradual slope (ideally 10:1 or greater slope; horizontal distance: vertical change), some fill material and grading may be required. In other states such as Virginia, fill is recommended to create a shoreline/beach behind a detached breakwater/oyster reef. This is not standard practice in Florida at the time of this writing, and in known cases such as Cedar Key, Florida, where it has been done, it has required more

FIGURE 7.2 *(A)* During the first stages of the project, team members will assess the shoreline, and then site work can begin. *(B)* This shoreline in Cedar Key, Florida, required extensive site work, but most locations do not require this level of intervention. *UF/IFAS file photos by Tyler Jones.*

permitting scrutiny. However, some hybrid living shoreline designs may require fill behind gray structural elements, such as retaining walls, terraces, or mangrove planters to create planting areas. (See table 4.1 for examples of hybrid living shorelines.)

Invasive species removal is also part of the site preparation. The management of invasive species will be part of the ongoing maintenance, and this is a requirement for both federal and state permits for any shoreline work. Often the contractor or the environmental consultant/restoration professional will oversee this. Also, they can work with homeowners, shoreline managers, or landscapers to identify the invasive plants and to demonstrate how to remove them or treat stumps with herbicide so they do not readily grow back. This way the homeowners or landscapers can continue to remove invasives on their own to reduce the likelihood of these species becoming established or reestablished. Local extension agents, members of local native plant society chapters, or landscapers can help to identify invasive plants as well as other native and nonnative plants.

## Step 2. Install Sill, Reef, or Breakwater

If the living shoreline plan includes an oyster reef or breakwater, these components should be installed first, because installation activities such as foot traffic or use of heavy equipment could damage shoreline plants. Also, since these hardscaping

## INVASIVE PLANTS COMMON TO FLORIDA'S COASTAL AREAS

Most of the nonnative plants growing in Florida are not invasive. Invasive plants are nonnative plants that have been shown to replace or damage native ecosystems. This is determined after years of study and observation by ecologists and other biologists working with the Florida Invasive Species Council. At the time of this writing, according to the invasive list on the Florida Invasive Species Council website, there are about 170 invasive plant species in Florida. (For more details on invasive plants, see chapter 2.)

Even if a known invasive plant does not appear to be aggressive in homeowners' yards or on other properties, it still should be removed and disposed of in the trash, not in the yard waste, so that it will not continue to spread.

Invasive shoreline plants in Florida that are commonly observed include the following:

- Australian pine (*Casuarina equisetifolia*)
- Beach vitex (*Vitex rotundifolia*) especially in Florida's Panhandle
- Brazilian pepper (*Schinus terebinthifolia*) in Central and South Florida
- Beach naupaka (*Scaevola taccada*)
- Torpedo grass (*Panicum repens*)
- Wedelia (*Sphagneticola trilobata*)

The Florida Invasive Species Council (https://www.floridainvasives.org/) is a 501(c3) nonprofit whose mission is to "reduce the impacts of invasive plants in Florida through the exchange of scientific, educational, and technical information."

The Center for Aquatic and Invasive Plants (https://plants.ifas.ufl.edu/), a department of the University Florida's Institute Food and Agricultural Sciences (UF/IFAS), focuses on the identification and management of invasive species.

### RESOURCE

Iannone, Basil V., et al. "Standardized Invasive Species Terminology for Effective Education Outreach." Publication FOR370, FR439. *EDIS* 4 (2021). https://edis.ifas.ufl.edu/publication/FR439.

elements protect living shoreline plants from wave or wake energy, they should be installed before the plants.

Some living shoreline projects use a phased approach, where reef or breakwater elements are installed weeks or months before any fill or plantings. Phasing in a project slowly has several benefits. First, time allows observations of the performance of the reef elements in different tidal and wind conditions. If observations indicate that wave energy is not sufficiently mitigated, corrective steps can be taken before funds and effort are expended on planting. Second, if reefs are reducing wave energy and sediment supply in the area sufficiently, material may begin to build up behind the breakwaters/reefs and could serve as an excellent substrate

**FIGURE 7.3** Photo panel of common invasive shoreline plants. *(A)* Australian pine (*Casuarina equisetifolia*); *(B)* Beach vitex (*Vitex rotundifolia*); *(C)* Brazilian pepper (*Schinus terebinthifolia*); *(D)* Beach naupaka (*Scaevola taccada*); *(E)* Torpedo grass (*Panicum repens*); *(F)* Wedelia (*Sphagneticola trilobata*). *Photos courtesy of the UF/IFAS Center for Aquatic and Invasive Plants.*

for future plantings. Third, a phased approach can help spread out project costs over a longer duration, which may be attractive to some property owners. But a phased approach may not always be allowable, depending on the permit conditions or project design. For example, a phased approach may not be appropriate for sill or planter designs, because the sill/planter function is directly tied to the presence of vegetation behind it.

Deployment of breakwater materials from land is less expensive, since it avoids the logistics of vessel use. Smaller structures like sills that use fiber logs or smaller reef modules (shell bags, COREs, oyster gabions, etc.) can be hand-placed, which allows homeowners or volunteers to be directly involved. If a water-based deployment is necessary due to the weight and large size of the rocks or reef modules and the inaccessibility of the shoreline, the team will handle the logistics of placement.

For oyster reefs and sills, there are a multitude of material types that provide substrate and/or contain shells (so they do not get washed away by waves and currents). Likewise, there are many choices for breakwater materials such as natural rock and prefabricated concrete modules. (See chapter 5.) These may be stacked in a series of curved mounds with gaps in between to allow water, people, and

**FIGURE 7.4** Oyster reef elements can be installed in various ways. *(A)* Larger reef elements are typically installed during high tide using a water-based approach, often using heavy equipment such as a barge with a crane (*photo by Reef Innovations Inc.*). *(B)* Oyster reefs and sills built out of smaller materials such as reef prisms are often deployed by hand at low tide, using a land-based approach *(UF/IFAS file photo by Tyler Jones)*.

marine life access through the reef. Oyster shells or other substrates should be installed in shallow water, so they are exposed to the air at low tide. Recent research has shown that oysters thrive within a specific tidal range (generally 6 to 18 inches [16 to 46 cm]) below mean sea level, or the average of sea levels over all tides, measured over an extended period at a particular location. Oysters grow best when exposed to the air 20 to 40 percent of a daily tidal cycle (two low tides and two high tides).[1]

Techniques have been developed using precise surveying instruments to fine-tune the placement of reef structures to take advantage of the tidal heights and oyster requirements. Members of the restoration team may have access to these tools, but even matching the elevations of new reefs to any existing oysters on existing natural (shells, rocks, or red mangrove prop roots) and man-made structures (dock pilings or concrete rubble) should be more than adequate to achieve proper reef placement. (See more details in chapter 6.) Aside from placement at an optimal tidal height, there may be regulatory limits to the height of an oyster reef breakwater. For example, the Florida General Permit for Low Profile Oyster Habitat Restoration limits the overall height of oyster reefs to 18 inches (0.5 m). (See section 62–330.632 in the Florida Administrative Code.)

### NON-OYSTER BREAKWATERS

If the site is in an area that will not support self-sustaining oyster habitat because of low salinity or other environmental conditions, a living shoreline breakwater can still be constructed to provide protection from moderate to high waves. In this situation, the breakwater should be constructed from materials other than oyster shell, such as coquina or limestone rocks, since oyster shells are in limited supply and should be reserved for areas suitable for oyster establishment. Although the structure will not serve as oyster habitat, it will still provide the primary benefit of shoreline protection and habitat for fish and other organisms adapted to local conditions.

Depending on the site, structures may still be required to be submerged at high tide. The permitting process will ensure that structures will not be placed in areas that impede boating activities.

### BUILDING A HYBRID LIVING SHORELINE

If there is already a seawall or riprap revetment, or if the site is on a canal, the process will be similar to the steps outlined above, with some exceptions. Green, nature-based enhancements to man-made structures will result in a hybrid living shoreline retrofit.

Hybrid living shoreline systems use natural materials like plants and oyster reefs in combination with hard elements like riprap (strategically placed piles of natural coquina stone or recycled concrete chunks). Nature-based elements can also be used to create a living shoreline retrofit in an area where there are existing gray structures like riprap revetments, bulkheads, or seawalls. Gray and green infrastructure may also be constructed together to create a hybrid living shoreline from the beginning rather than retrofitting existing structures.

**FIGURE 7.5** *(A)* Waterward edge protection can take the form of breakwater reefs, where the reef elements are separated from the vegetated area (*photo by Allison McDowell, Choctawhatchee Basin Alliance*); or *(B)* sills, where the oyster colonization substrate (reef prisms) is placed directly adjacent to the vegetated zone *(photo by Savanna Barry)*.

Examples of these may be retaining walls or terraces. A mangrove planter is an example of a constructed hybrid living shoreline component built to create a graded planting area in front of an existing vertical seawall. (See chapter 4 for examples.) Since they include the incorporation of traditional gray infrastructure, hybrid living shorelines usually require more engineering and the use of heavier construction materials and techniques; and as a result, they will be more expensive than fully nature-based living shoreline options.

Hybrid living shorelines are used in areas where wave energy is too high for a nature-based shoreline. Hybrid shorelines are suitable for areas with medium energy, boat wakes, or a steeper slope. They are also used in areas where homeowners

**FIGURE 7.6** Various reef-building materials are available that can be used to create viable oyster habitat without introducing plastic to the marine environment. *(A)* Some, like this reef ball, are small enough to be easily carried. *(B)* Others are modular, like reef castles which stack and interlock to form a larger structure. *Photos by Savanna Barry.*

want to add natural elements to an existing hard structure to increase habitat, shore up a failing structure, increase the lifespan of a hard structure, or for aesthetic reasons.

Riprap may be added to the shore to provide a higher resistance of the bank to wave action. Coquina and limestone are recommended because they are natural to Florida and provide habitat for plants and animals, including oysters. Recycled chunks of cleaned concrete can also be used as riprap. Generally, riprap is placed inshore of oyster reefs and intertidal plants. It can also be added at the base of an existing seawall or bulkhead to increase substrate complexity and decrease scour or erosion that occurs at the bottom of the seawall.

**FIGURE 7.7** The armored shoreline at the Nature Coast Biological Station in Cedar Key, Florida, was enhanced by adding reef balls outside of the riprap to create additional habitat. *Photo by Jim McFarlane.*

## OPTIONS FOR CANAL-FRONT PROPERTIES

Florida has thousands of miles of canals that have been dug to provide more waterfront properties, and most of these have seawalls. (See chapter 2.) They present a distinct problem, especially where lawns extend to the edge of the seawall, which allows lawn chemicals and stormwater carrying pesticides, organic materials, and fertilizers to enter the body of water. In addition, the flat structure of a seawall offers almost no habitat.

Canal-front homes usually offer little space for living shorelines, and their seawalls create abrupt drop-offs that eliminate the natural connection between land and water. However, canal-front homeowners can enjoy some of the benefits of living shorelines by retrofitting seawalls with riprap and mangroves. Coquina stone riprap placed at the base of the seawall helps prevent scouring, which undermines the seawall. Natural stone can also be used to create mangrove planters. Adding these elements to seawalls prolongs their life and offsets some of the habitat loss. Mangrove wall inserts or other seawall modifications can also be made. Sculpted panels attached to the face of the seawall are aesthetically pleasing and increase substrate complexity by creating nooks small creatures can hide in.

By adding these features, even a canal-front shoreline can be transformed into habitat. In addition, installing an upland buffer of native plants will improve the water quality. (See chapter 8.)

## Step 3. Install Plants

Although plant installations are shown here as step 3, this may be the next step after preparing the site if a sill or breakwater structure is not required. Building the hard, more waterward structures first is recommended, if that is part of the design, because it will be easier to install plants after those hard structures are in place. This will also be necessary if fill is added to create planting areas in mangrove planters, terraces, or sills.

Smooth cordgrass, rushes, and mangroves are the native plants most often used in the intertidal zone in South and Central Florida. In the northern part of the state, mangroves will probably not be used. Since these salt marsh plants spread via rhizomes, they stabilize the land-water interface at the low- to mid-intertidal zones. Once these plants are established, then a wider diversity of plants can be installed.

The team, homeowners, or shoreline managers should try to source only locally grown plants for the site. The Florida Living Shorelines website (https://floridalivingshorelines.com) is a good resource that has not only suggestions on what to plant but also information on the proper tidal zonation for various species and on which plants are appropriate by region. All plants listed are available commercially. As part of planning the project (See chapter 6.), the existing native plants on the shoreline should have been identified. Leaving these to grow and proliferate will contribute to the stabilization and biodiversity of the living shoreline. Additionally, these existing plants may be species that might not be available commercially or that might be more difficult to source. Whenever possible, integrate the installed vegetation with existing native vegetation to improve the shoreline's ecological function.

### GRADE AND ELEVATION ZONES FOR MARSHES AND MANGROVES

Ideally, shoreline elevations can be tied to surveyed elevations referenced to local tidal datums or benchmarks for the site, for example, NAVD88 is a commonly referenced tidal datum for Florida and the East Coast of the United States. This referencing may be done with surveyor's equipment that can measure geo-referenced elevations. A professional surveyor, or a member of the restoration team familiar

**FIGURE 7.8** Tidal wrack is the dead material that washes up on shorelines with each high tide. The "wrack lines" left behind are useful natural clues as to level of the typical high tide line. By marking the location of wrack lines after normal high tides (*A*: yellow and orange lines in example photo) and larger storm events (*B*: red lines in example photo), property owners can gain a good understanding of the location of the mean high water (MHW) line without a formal survey. This information is important for planning locations of oyster reefs and planting zones and may be required information for permitting purposes. *Photo by Larry Beggs, Reef Innovations Inc.*

with these methods (and who has the equipment) can perform this service, if the project budget allows. If there is a nearby shoreline or marsh site that can be surveyed, the elevation data obtained from this natural reference site can be used as a template to guide where planting zones can be established. If possible, a comparison of a reference site with the planned project site would be the ideal approach.

If a professional survey is not in the budget, other indicators can be used to infer intertidal planting zones. Noting the location of wrack lines can be a means to delineate approximate tidal zones. A wrack line or zone is the area on the shoreline where organic material and other debris are deposited at a high-water line. The regular high tide wrack line that is seen on average, or normal, high tides generally delineates the mean high water (MHW) line, or the average of typical high tides. Wrack lines left behind by higher-than-normal tides, such as spring or king tides, or by storm events roughly demarcate the mean higher-high water (MHHW) line, or the average of the highest water levels. Marking these wrack lines will help designate high marsh (between MHW and MHHW), and low marsh (below the MHW to water line at the average low tide line, or mean low water [MLW] line).

Another approach would be to visit the site at peak high tide and trace the water line using a handheld GPS unit. Several repeat visits would be necessary to map the approximate MHW. This can be repeated during several king tide events to map the MHHW. These GPS locations can be overlaid on aerial images using GIS software such as ArcGIS or Google Earth. The lines marking the MHW and MHHW on imagery can be used to prepare site plans for permitting and planning planting.

Existing native shoreline plants on adjacent shorelines can indicate where plants on the project site should be installed. If the stretch of shoreline encompassing the project site and adjacent locations is relatively straight, this helps give a better approximation of where installed plants should go. This is still an approximation, as elevation and shoreline configurations may vary.

### GRADING OR TERRACING

The slope of the shoreline can be characterized by the ratio of the run (horizontal distance) to the rise (vertical change), where smaller ratios indicate steeper slopes. When slopes are very steep, such as 3:1, grading the shoreline to a gentler slope, ideally 10:1 or lower, can result in a wider and more successful planting area. This may involve heavy equipment, which increases the costs. Careful consideration during the planning stages is necessary to plan for these costs and to include use of this equipment in permit applications.

If grading is not practical, then terracing may be another approach to gradually decrease the overall slope steepness over a series of platforms. The platforms are reinforced and retained with wood, rock, or concrete like bulkhead construction. The elevation of each terrace represents different planting zones with corresponding species suited to each elevation.

### PLANTING ZONES

Moving from the water inland, appropriate marsh or mangrove species will change accordingly. Three main elevation zones for planting are low marsh, high marsh, and coastal upland. The low marsh zone is from the MHW to mean sea level at low tide, and it is flooded daily at high tide and exposed at low tide. Smooth cordgrass is appropriate for this zone. Red mangroves are also appropriate for the low marsh areas, although the planting of propagules or small sizes, for practical purposes, should be done around the MHW or just above MHW (see chapter 5) to reduce the chances of washouts during higher tides and waves. Mangroves, when successful, will migrate out farther into the low marsh zone. Smooth cordgrass plantings can also facilitate the recruitment of mangrove propagules by trapping them and allowing them to root and establish. Black mangroves should be planted just landward of red mangroves and can traverse the transition from low to high marsh.

The high marsh zone is between MHW and MHHW. Plants at the MHW or just above get inundated at high tides, while plants above the MHHW get flooded only on the king tides or high-water storm events. Plants in the higher parts of this zone generally stay dry. Appropriate high marsh plants include black needle rush and salt marsh hay. Plants for the higher reaches of this zone include saltgrass and seashore paspalum. White mangroves are also appropriate in the upper high marsh zone.

The coastal upland zone does not typically get wet or inundated except during extreme events. Sea oats, panic grass, seaside goldenrod, muhly grass, dune sunflowers, sea oxeye daisy, wax myrtle, and green buttonwood are some examples of plants that can be installed in the coastal upland zone. Increasing the number of species that are planted helps to enhance biodiversity of a living shoreline. For the names and descriptions of the range of plant species that can be used in each zone, see appendix A.

## PLANT DEPTH AND SPACING

Generally, planting depths of 6–8 inches (15–20 cm) should anchor the plants enough to avoid uprooting due to tidal currents and waves, while ensuring they do not get too wet and potentially smothered. Mangroves should be planted 1–2 inches (2.5–5.0 cm) deeper than they are potted. This will allow sufficient aerial exposure of their prop roots or pneumatophores for oxygen delivery to buried roots. Anchoring young plants by tying them to a metal or fiberglass rod may be necessary if there are concerns in a higher-energy environment. Holes should be filled, and plants should be snug in the ground to minimize uprooting.

Traditionally, grasses are planted 1.5 to 3 feet (0.5 to 0.9 m) apart in a staggered pattern, and mangroves are planted farther apart (2 to 8 feet [0.6 to 2.4 m]) depending on the size of the planting unit) to accommodate their spread. This dispersed design is best suited to low-stress environments because it reduces competition between transplants.[2] Grasses may look sparse at first, but once they are established, they will spread and fill in. By contrast, in more stressful environments, denser initial plantings of grasses, and/or planting multiple stems per hole will help to establish a marsh habitat in less time because the plants help support

**Planting Patterns**

**FIGURE 7.9** Images of planting patterns for marsh plants in intertidal zones. *(1)* Traditional single evenly spaced plant plugs planted 18 to 36 inches (0.5 to 0.9 m) apart. *(2)* Checkerboard/sod plantings with 10-inch (25 cm) squares. *(3)* Evenly spaced 10-inch (25 cm) square clumps in the middle of 6.5-foot (2 m) squares. *Illustration by Marjorie Shropshire.*

each other when they remain in contact. In fact, research has shown that planting intact clumps (i.e., sods) of marsh species in a checkerboard pattern can lead to rapid establishment.[3] At the time of this writing, sods of marsh species are generally not available commercially, making it difficult to apply this practice broadly in living shorelines. Other research shows that installing plug transplants in a clumped design can also increase plant yields and lead to faster establishment than the traditional dispersed design.[4] Despite research on sod-based and clumped designs, the vast majority of planting designs still employ the traditional, dispersed design. Work with your team to determine the most appropriate planting plan based on your site, plant material available, and current scientific recommendations. Marsh grasses and rushes hold sediment in place and can also trap seeds of other marsh plants, allowing a more diverse natural plant community to develop and grow.

### TIMING OF PLANTING

Plants will do best when installed in the spring (March to May); earlier is better especially in South Florida. This allows plants enough time to establish and spread roots through the summer before hurricane season peak (September), when hurricanes and other seasonal high-water events typically occur. Planting in fall or late winter can be done as well, but many plants may be dormant during this time and may be slower to self-anchor and spread, and they may also require irrigation to survive through the dry season. The farther south one is in Florida, the earlier one is able to plant. Care should be taken to avoid planting during high-water periods. Even during optimal months, plantings should be done at low tide with attention to the tidally based planting zones previously mentioned.

#### PLANT CONDITION AND SIZE

A general rule of thumb for living shoreline plants is to purchase the largest plants possible that the project budget will allow. The larger plant size means that they will become more resilient more quickly and withstand wave action on the site. This is especially important for mangrove species, particularly red mangroves, as they will likely encounter the most frequent wave action and tidal inundation throughout the year. (*Note:* This advice is the opposite of that for terrestrial plants, where smaller, younger plants are recommended because of their ability to quickly adjust to their new planting sites.)

Since most nursery-grown plants are grown using fresh water, it is a good idea to acclimate the plants to the salinities they will eventually experience at the project site. Acclimate low marsh plants and black mangrove trees to salt water that is 10–15 ppt to prepare them for most salty conditions. To avoid shocking the plants with a saltwater irrigation regimen, use a stepwise increase of salinity at 5 ppt per week until the water is within 5 ppt of site salinity or reached entirely. Red mangroves, however, do not need to be acclimated as they are salt excluders and can go directly from freshwater nursery conditions to saltwater environments. For a summary of planting considerations for living shorelines see Florida Sea Grant's video, *Planting Considerations for Living Shorelines* (2021), which is available on YouTube (https://youtu.be/2jf6w8DOzOs?si=IatnHT4W7uZIvQTP).

### Step 4. Monitor Progress

Once the elements of the shoreline are installed, the homeowners, shoreline managers, or team members should check on it during the establishment period. Any of the installed plants that die should be replaced. The site should also be monitored regularly to record the growth of grass and oysters, to determine whether sand is accumulating, and to make sure no erosion is occurring. Additional monitoring of the shoreline should also be scheduled after storms. Make sure to discuss the monitoring schedule with the team. As part of the permitting process, monitoring may be required for at least one year to determine whether the project is successful. Generally, teams should schedule more frequent visits just after installation and then gradually increase the time between visits. For example, a typical

**FIGURE 7.10** *(A)* Nursery-grown 1-gallon smooth cordgrasses were planted on Joe Rains Beach in Cedar Key, Florida. Larger-sized plants were chosen for this site since they could withstand wave action better than smaller plants (*UF/IFAS file photo by Kacey Aukema*). *(B)* The Snook Islands project in southeastern Florida incorporates red mangroves and smooth cordgrass. Grasses are planted in the shallower parts of a created sandbar, while the mangroves are planted in deeper water *(photo by Ginny Stibolt)*.

schedule might be to check on a project weekly for the first month, monthly for the next three to six months, quarterly for two to three years, and then annually thereafter.

The homeowners can also keep an eye on their shorelines to see what kinds of wildlife are visiting. It is a good idea to take photos as the shoreline matures, since sharing success with others is good outreach. The growth and health of the living shoreline will reflect the benefits it is providing to the waterfront. Sediment accumulation and the spreading of plants indicate that the shoreline is getting wider and therefore providing more protection to the waterfront. Oysters may be growing and increasing in number with more spat settlement. (See chapter 1.) As oysters thrive as part of a living shoreline, they will provide not only protection but also more filtration to improve water quality. Although it may be difficult to see an improvement in water quality, an increase in wildlife such as fish, crabs, and birds visiting the shoreline and taking advantage of the increased marsh or mangrove as well as underwater habitats can be an indication of not only improved habitat but also improved water quality.

Below are some tools and metrics used to measure and track the progress and health of the living shoreline project. There will be some required measurements as part of the permit, but there may be additional ones that are easy to perform and that can provide more information on the project's success. Monitoring in addition to that required by the permit may be welcomed by permitting agencies, but homeowners could incur additional costs for these services from members of the professional team. Homeowners or site managers are encouraged to carry out as much monitoring as is feasible, especially because more data can inform and improve future living shoreline practices. All the methods described can be performed by more experienced project team members, but with training, homeowners and other property managers can participate in the more advanced measurements, too.

Several of the measurements described below can be performed before installation, as part of the site assessment; and if relevant, these measurements may be included in the permit application documents. After installation, the same techniques can be repeated to compare post-installation measurements with pre-installation measurements. Details on these measurements are provided below, but also see Laura Reynolds and colleagues "Living Shoreline Monitoring."[5]

## NEEDED SUPPLIES AND TOOLS

**QUADRAT** A quadrat is a portable frame of a known area—typically a square but it can be circular as well—that ecologists use to delineate a sampling zone within which data such as stem counts of plants and counts of stationary, or sessile, animals can be recorded. These counts allow for calculations of species' densities and for estimating the percentage coverage of an area within the living shoreline by a particular species. When repeated on a regular basis, this provides a measure of how well the shoreline is adapting; it also makes visible the complexity of the developing ecosystem. A quadrat can be easily constructed from plastic PVC pipe (3/4 in. [19 mm] diameter is recommended) into a 0.5 × 0.5 meter (1.6 × 1.6 ft.) square. Quadrat evaluation of plants and oysters is typically associated with more advanced monitoring methods, but the layperson can readily learn to use one or at least photograph quadrats that can then be analyzed by other project team members.

**RULER OR METER STICK** Small rulers (6 in. and 12 in. [15 cm and 30 cm]) and a meter stick are useful for measuring the lengths and heights of animals (such as oysters) and plants.

**PVC SEDIMENT CORER AND MESH SIEVES** A simple sediment-coring device can be made from an 18-inch (0.5 m) length of 3-inch (0.1 m) diameter PVC with an endcap. A sieve for catching infaunal organisms (invertebrate aquatic animals that burrow in the sediment, such as clams and worms) and separating out sediment can be made from materials that can be purchased at a hardware store such as window screen, hardware cloth, and chicken wire. Square-shaped mesh works best. The team of professionals may construct sieves with wooden frames or purchase a set from an environmental supply company. Analyses of sediments using these devices are typically associated with advanced monitoring methods.

**NETS AND TRAPS FOR FISH** Minnows to bait fish traps can be purchased from a bait and tackle shop, outdoor sports retailer, or wildlife/environmental supply company. The traps can also be baited with frozen bait purchased from a bait and tackle shop. Cast nets can be purchased from bait and tackle shops or other fishing/outdoor sports retailers. For the homeowner who is also an avid fisher, a rod and reel can be a valuable monitoring tool that provides enjoyment (and maybe dinner!) as well. (See the section titled "Basic Monitoring for the Homeowner and Other Property Managers" below.)

**BUTTERFLY OR SWEEP NETS** Butterfly or sweep nets can be purchased from most retailers that sell wildlife and camping/outdoor supplies. Larger nets for environmental sampling are recommended, but kid-size nets can be purchased as a fun way to involve children in monitoring the living shoreline.

**FIGURE 7.11** A variety of monitoring supplies such as quadrats, sediment corers, and seine nets can be used to collect information. *(A)* Baited minnow traps. *(B)* Coring tube. *(C)* Dip net and sieve. *(D)* Seine net. *Photo A by Savanna Barry. Photos B, C, and D, UF/IFAS file photos by Tyler Jones.*

**BINOCULARS** Sturdy, outdoor binoculars for viewing birds and other wildlife can be found and purchased from a variety of retailers—either in-person or online. We encourage speaking to local birding groups such as a local Audubon chapter to gain any specific product recommendations.

**FIELD GUIDES FOR SPECIES IDENTIFICATION** There are several field guides that can be purchased that can help identify most plants and animals that occur in shoreline habitats. Inquire at the local extension office and at local environmental nonprofits about identification resources more specific to the area. After becoming familiar with the flora and fauna on the living shoreline, it could be helpful for

**FIGURE 7.12** Photos can be used to document animals visiting a living shoreline. *(A)* American Oystercatchers. *(B)* Osprey. *(C)* Wood Duck. *(D)* Bald Eagle eating a fish. *Photo A by Florida FWC/Karen Parker, 2015, Flickr (https://www.flickr.com/photos/myfwcmedia/24394059022/in/album-72177720304442295; reproduced with cropping under creative commons license CC by 2.0, https://creativecommons.org/licenses/by/2.0/). Photo B, UF/IFAS file photo by Tyler Jones. Photos C and D by Ellyn Stevenson.*

homeowners to develop their own field guide by keeping a wildlife diary. Wildlife apps can help identify new species by sharing photos with the community of app users. (See Cameras and Smartphone Apps below.) Crowdsourcing identifications can be a fun way of learning new species.

**NOTEBOOK OR DATA SHEETS** Most any notebook will do, but one that has waterproof paper will allow bringing it outside under a wide range of weather conditions. Recording observations of the living shoreline wildlife through writings and drawings could serve as a great way to get started in nature journaling as a hobby.

**CAMERAS AND SMARTPHONE APPS** A time series of photos will always be a useful record of the living shoreline's progress for required monitoring. Smartphone cameras are more than adequate. Shoreline changes from a cross-sectional perspective can be photographed from fixed locations that can be revisited. It is a good idea to revisit the same location and take a photograph at regular intervals, such as monthly or quarterly, at the same tidal stage (ideally low tide) to best facilitate comparisons over time (and do not forget to take a good set of "before" photos, too). Fixed locations can be marked with poles or a platform on which to mount a camera. It is a good idea to add weather/waterproof cases for any camera used to record living shoreline progress or to use cameras specifically geared toward outdoor use. Specialized wildlife cameras can also be a unique way to capture animals visiting the living shoreline, especially nocturnal species. There are several manufacturers and models that are available for the hobbyist at economical prices, and they are easy to set up.

One waterfront homeowner with a living shoreline (see Case Study #3 in chapter 6) enthusiastically documents the wildlife that now visits her shoreline. In addition to her personal camera, she mounted a wildlife camera with nighttime vision, which captured various nocturnal visitors. Animals visiting this living shoreline area included coyotes, bobcats, river otters, and multiple bird species.

If using a smartphone as a camera, several smartphone applications make recording wildlife an interesting family activity. A useful and fun app that can be used to monitor plants and wildlife is iNaturalist (https://www.inaturalist.org). There are other apps specifically geared toward birding such as eBird, Audubon Bird Guide, or Merlin Bird ID. Fish catches can be recorded with iNaturalist or with specific fishing apps such as FishBrain. There are many others as well.

## BASIC MONITORING FOR THE HOMEOWNER AND OTHER PROPERTY MANAGERS

The following data collection methods are ones that nonprofessionals can easily learn and conduct either on their own or with their project team. Some methods even lend themselves to the involvement of family and friends, providing fun outdoor activities while collecting data on the project's progress.

SHORELINE GROWTH The simplest method to record changes in shoreline growth is to photograph changes over time from a fixed location. At least one point should capture a cross section, or side view, of the living shoreline to show the growth of plants and changes in the width of the shoreline. Photos can be taken at the same frequency as transect measurements, at least twice per year, and after significant storm events.

Photos capture the visual impact of the project's success and can be used by the team as a selling point for future projects. These photos can also be used in outreach, to convince neighbors, communities, municipalities and other shoreline managers of the success and value of installing living shorelines to protect their waterfront properties. The photos also augment quantitative measurements of shoreline growth. (See the section titled "Shoreline Growth" under "Advanced Monitoring Methods" below.) Photographic evidence, coupled with more precise measurements, can have a tremendous impact in demonstrating the resiliency of a living shoreline.

PLANTS Photos provide a simple, impactful way for the homeowner to document how plants grow and develop on a living shoreline. The photos used to record changes in shoreline growth will also capture how plants change at a landscape level. Biodiversity can be recorded by photographing individual species, both those installed and new recruits from the wild. These photos can then be uploaded into a citizen science app such as iNaturalist. An added benefit of iNaturalist is that the larger community using the app can help identify unknown species, providing a fantastic learning opportunity for the novice.

Homeowners can also assist in more advanced, quantitative methods of quadrat sampling by simply photographing quadrats of fixed locations over time and providing the photos to project team members responsible for these measurements. For more explanation on advanced methods to record changes in plant growth and diversity, see the section "Advanced Monitoring Methods," below.

**FIGURE 7.13** Mangroves are important for the continued presence of sportfish, like this redfish *(Sciaenops ocellatus)*, which was caught and released by Capt. Alex Gorichky near a living shoreline with a robust mangrove forest as seen in the photo. *Photo by Logan Gorichky.*

ANIMALS Recreational fishing is a fun way to determine whether larger fish are using the living shoreline habitats. Traditional fishing gear such as rods and reels or cast nets can be used. Homeowners may even be able to catch crabs around their living shorelines with crab pots, if conditions are suitable. Again, keep in mind the local and state fishing regulations for catch sizes, seasons, bag limits, and gear. Photos of catches may be some of the most impactful evidence supporting the value of living shorelines as a practice to make waterfront properties more resilient and ecologically friendly. The photos can be uploaded to a fishing or wildlife app to keep track of the catches and help identify them if needed. (See examples in the section titled "Cameras and Smartphone Apps" above.) The Florida Fish and Wildlife Commission's Catch a Florida Memory (https://catchafloridamemory.com/) program allows fishers to submit photos of their catches and win prizes for their entries. If homeowners are fishing enthusiasts, they can, in pursuing their hobby, collect valuable data to document how their living shoreline serves as good fish habitat while enjoying years of good fishing as the living shoreline evolves and grows.

**BIRD AND BUTTERFLY WATCHING** Monitoring birds that may start visiting and using the living shoreline will provide years of enjoyment. Homeowners will need only a good set of binoculars, a field notebook, a camera, and possibly a few key smartphone apps to upload the sightings or even record bird calls for identification. For scientific monitoring purposes, observations should be performed for a set period (fifteen to thirty minutes for example).[6] Note the conditions when making these observations (date/season, time of day, tide, weather conditions, etc.). The types of species and abundances will vary according to conditions. There are a variety of coastal birds that are likely to be seen using living shorelines as habitat. As the living shoreline becomes more established, those observations in both diversity and abundance are likely to increase. Bird diversity and use of living shoreline projects may be valuable data for state agencies like the Florida Department of Environmental Protection (FDEP), especially if the waterfront is in or near an aquatic preserve.

For the homeowner, noting the presence alone of bird species visiting the living shoreline, accompanied by photos, still serves as valuable data. Photos can be uploaded into apps like iNaturalist. These living shoreline observations can also provide data for ongoing citizen science efforts such as the Audubon Christmas Bird Count (https://www.audubon.org/community-science/christmas-bird-count).

Monitoring butterflies and other flying insects can be accomplished by photographing any notable or unique species. Again, these photos can be uploaded into iNaturalist or similar apps.

**OYSTERS** If oyster reefs were installed as part of the living shoreline sills or breakwaters, then the monitoring of oysters should be performed. Scientific methods are described in the following section. For the homeowner, noting oysters' appearance and sustained presence via sequential photos acts as a useful complement to these methods. A resident has the advantage of noticing these changes firsthand, which can trigger the start of monitoring efforts. Observations alone are powerful tools that aid the project team in their monitoring efforts. As with plants, the homeowner can also photograph quadrats to assist the team. Photos of oysters and other species that will start to occupy the new oyster habitat can also be uploaded to iNaturalist for identification and documentation.

## ADVANCED MONITORING METHODS

**SHORELINE GROWTH** A long-term transect can be established through the living shoreline to track changes in the width of the shoreline. Permanent poles can mark the transect line, which should traverse the planting area and extend out to any breakwater structures. The distance from a permanent and fixed landward

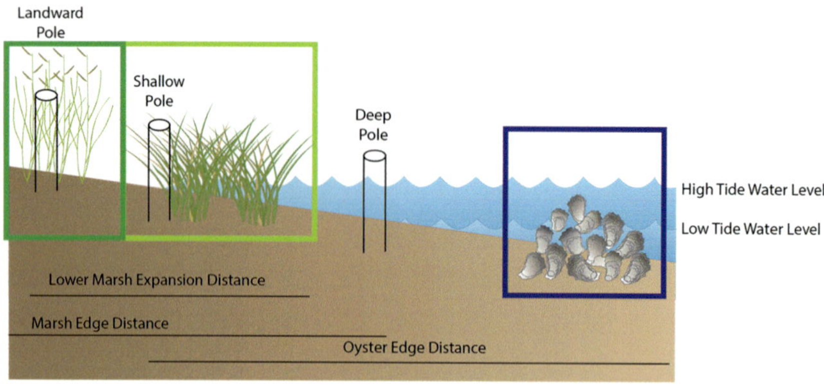


**FIGURE 7.14** An example of a transect that will assist with measuring the expansion of a living shoreline (or accretion). The poles are the permanent markers, and the lines in the figure represent the distances to measure. The dark-green box indicates high marsh; the light-green box indicates low marsh; and the blue box indicates the oyster region. *Illustration reproduced with permission from Laura Reynolds, Natalie C. Stephens, Savanna C. Barry, and Ashley R. Smyth, "Living Shoreline Monitoring—How Do I Evaluate the Environmental Benefits of My Living Shoreline? SS694/SL481, 1/2021," EDIS 2021, no. (2021), https://doi.org/10.32473/edis-ss694-2021. Icons by Tracey Saxby, Integration and Application Network (ian.umces.edu/media-library; licensed under Attribution-ShareAlike 4.0 International [CC BY-SA 4.0]).*

pole to the outermost extent of the living shoreline plants ("marsh edge distance") can be measured periodically over time (at least twice per year) and after any major storm events to determine whether the shoreline is growing and is providing protection. In addition to multiple measurements over time, it is ideal to have several points at which transect measurements are made at each data collection event. The team can establish these points and perform the measurements or train the homeowners or shoreline managers to measure them.

An increase in marsh edge distance represents a widening of the living shoreline as sediment accumulates and as plants migrate waterward. Also measuring the distance between the MHW line and the MLW line provides a representation of intertidal width.[7]

Depending on the shoreline's configuration, intertidal width may increase as the living shoreline grows waterward. It may be difficult to measure this by eye, however, especially as tide levels vary depending on the phases of the moon that affect the tidal amplitude. Fixed landmarks such as installed poles, preexisting vegetation, and man-made structures can help in making these measurements. And repeated measurements are necessary to provide a more complete analysis of the shoreline.

Intertidal width can also be monitored by establishing a long-term transect across the living shoreline, through the high and low marsh, to the oyster reef or breakwater structure. Marsh expansion can be measured from a fixed upland pole (transect origin) to the waterward edge of each of the marsh vegetation zones and the oyster reef. Use the additional poles (e.g., shallow pole, deep pole) to line up the transect to ensure that measurements are performed along the same straight-line distance over time. The distance from the fixed upland pole to outer vegetation zone edges will increase as the marsh grows waterward, representing an expansion or increasing width of the shoreline (this is also called the lower marsh expansion distance).

**SEDIMENT CHANGES** After the living shoreline structures are installed, different types of sediments may accumulate behind or around them. Sediments vary according to size and color. For example, sand is light in color and has larger grain sizes relative to silt or mud, which will be finer and darker. As part of monitoring, noting these changes along with changes in the width or growth of the shoreline can be helpful. The team of professionals can do more in-depth monitoring, such as sediment coring to measure sediment sizes and types and to identify organisms that may be starting to colonize the sediments or benthic infauna (organisms living in the sediment). Briefly, coring is the process of collecting a standard plug of sediment usually using cylindrical sampling devices such as PVC tubes or modified syringes. Generally, a tube of a standard diameter is driven to a standard depth into the sediment, and the sample is extracted by creating suction on the top end of the tube using an airtight cap or plunger. Smaller diameters (e.g., 0.8–2 in. [ 2–5 cm]) and shallower depths (e.g., 2–4 in. [5–10 cm]) are usually used for sediment grain size/color analysis, while larger tube diameters (e.g., 2.4–4.8 in. [6–12 cm]) and deeper depths (e.g., 6–12 in. [15–30 cm]) are used for samples meant to examine invertebrates living in the sediments or plant community characteristics such as root biomass. Other methods to monitor the biological characteristics of living shorelines are described below.

**BIODIVERSITY AND DENSITY** Quadrat sampling is a common and effective method for measuring changes in plants within a living shoreline. All described quadrat measurements will be more easily performed at low tide. Place the quadrat on the seabed, over a section of the planted area. Identify and count all the plant species present within the quadrat. A starting count will be the number of plants by species that were originally installed. Although only a few species may have been planted as part of the living shoreline installation, over time, there may be different species recruiting into the planted area, which adds to the total species count and increases the biodiversity.

**FIGURE 7.15** Place a quadrat over a section of the newly planted area and identify and count all the plant species present. Using the same technique, monitor the site quarterly, seasonally, and after storm events to estimate the growth and diversity of shoreline plants. Document the quadrats with photographs to gauge success over time. *(A)* Example of sparse plant coverage (~25 percent) over the quadrat area. *(B)* Example of dense plant coverage (~100 percent) over the quadrat area. *Photos by Savanna Barry.*

To get an estimate of plant coverage, the number of plants by species can be divided by the area of the quadrat (which gives the number of plants per square meter). The percentage cover of plants can also be estimated within the quadrat. As the percentage of cover increases, it reflects the amount of lateral growth (or spreading of plants from their initial planting location), which is a good sign of the increasing growth and health of the living shoreline plants! Conversely, decreases in percentage cover could reflect burial or other damage from storm events or persistent wave action, or from other impacts to the health of shoreline plants.

At least three quadrat measurements should be made at each monitoring event, but more measurements are useful, if they are feasible for the site. Quadrats could also be randomly generated at each monitoring event, fixed within the planting area, or a combination of both. Quarterly or seasonal measurements and post-storm events should provide a good evaluation of the living shoreline plant populations.

**GROWTH AND SURVIVAL** In addition to the lateral growth of plants, measured by the percentage of cover within a quadrat, a meter stick can also be used to measure changes in height or the vertical growth of the plants. These measurements can be performed within each quadrat sampling and applied to individual trees such as planted mangroves. As mangroves grow taller, it might be simpler and more practical to track when they reach legal heights (6–10 ft. [1.8–3 m]) for trimming. (See height ranges for legal mangrove trimming figure 7.22.) A 10-foot PVC pole, the standard length sold in hardware stores, with a marking at 6 feet (1.8 m) can be used to estimate when mangroves can be legally trimmed. The pole can be marked with smaller increments (6–12 in. [15–30 cm]) to track growth up to 6 feet (1.8 m). Height should be made from the base of the prop roots at the sediment surface to the highest point in the canopy.[8] For smaller mangroves, canopy area can also be estimated from simple measurements along with height. A tape measure can be used to estimate canopy area by measuring maximum and minimum dimensions of the canopy ($\pi \times$ [max. canopy width $\times$ min. canopy width]). Canopy volume can be measured by measuring the depth, or height, of the leafy portion of the tree from top to base and multiplying the canopy height

**FIGURE 7.16** Plant community monitoring. Changes in plant species from the coastal uplands/high to low marsh can be documented along a transect tape line. Species and percent cover within a quadrat are recorded at regular, or random, intervals along the transect. Growth and survival of planted species and appearance of species recruited from the wild can also be noted. *Photo by Savanna Barry.*

by its area. When the mangroves become too large to measure, members of your team or other professionals trained in, and possessing the necessary equipment for measuring the trees, can continue monitoring the mangroves. A permanent, fixed photo station can be a great way of tracking how mangroves and other living shoreline vegetation grow and develop over time. (See the section "Cameras and Smartphone Apps" above.) A reference marker or pole of known length, or a friend can be used in the photo to give the scale to the vegetation. The vertical and horizontal growth of healthy vegetation in a living shoreline represents its growing resiliency to erosion and storm events.

Survival can also be easily measured from the observation of green leaves, but events such as high winds or coastal flooding may significantly defoliate a tree. The plants may recover and sprout new leaves, so check for brittle branches and scrape under the bark ("the bark test") to determine whether there is moist, living tissue underneath. This indicates the tree is still living. Keep in mind Florida mangrove regulations as well. Even dead mangroves are protected under Florida's 1996 Mangrove Trimming and Preservation Act and cannot be removed.

All these measurements should be accompanied by photos to document and track changes in growth and survival. There is no need to take photos of every quadrat or tree, but key photos can be representative of the changes and overall health of the living shoreline. The homeowners or shoreline managers and their team can use these photos to showcase the success of the project to neighbors and other waterfront property owners that may be unfamiliar with the benefits of living shorelines.

**MONITORING LIVING SHORELINE PLANTS AND ANIMALS** When using quadrats to measure plants in the marsh or mangrove areas, homeowners should also record the animals in this habitat. Over time, there may be marsh periwinkle snails, fiddler crabs, and other species. An increase in these species indicates a growing, healthy marsh. Sediment cores can be used to identify not only the types of sediments accumulating in the living shoreline but also the infaunal species colonizing the sediments. As with the quadrats, multiple cores should be taken to get a better estimate of biodiversity and density across the planted area. Identification and counts of associated species, along with the plant quadrat data, will provide a robust record of the biology of the living shoreline.

As mangroves develop prop roots, look for oysters and other sessile organisms attaching to the roots. The presence of tree crabs, spiders, and other animals are signs that the mangroves are becoming a viable, biodiverse habitat, all while protecting the shoreline. These animals should be recorded as well. Although it may be difficult to estimate densities on prop roots and in tree canopies, a count of different species is still a useful record of how the mangroves are contributing to the biodiversity of the living shoreline.

## OYSTERS AND OTHER ORGANISMS ON SILL OR BREAKWATER STRUCTURES

Quadrats can also be used to estimate biodiversity and density on shell or rock sills and breakwater structures. If the structure is large enough, the 0.5 × 0.5 meter quadrat used for measuring the vegetation can be used for this monitoring as well. The size of the structures, however, may be smaller, making it more difficult to use the same quadrat. Working on varying shaped and sized surfaces may require the construction of a smaller quadrat (for example a 0.25 × 0.25 meter square, or even a 0.10 × 0.10 meter square). The smaller and larger quadrats can be used together for varying scales, such as a broader percentage cover of oysters across a larger area and more detailed species counts in a smaller area.

Live oysters should be counted directly within each quadrat and density calculated as counts per square meter. The percentage of cover of live oysters can also be estimated, although care must be taken to distinguish newly settled oysters

**FIGURE 7.17** Oyster shell height is measured from the shell hinge to its farthest point of growth, or maximum length. A ruler or set of calipers can be used to measure the length of a live oyster. *UF/IFAS file photo by Tyler Jones.*

from existing shell, if the sill or breakwater structures are constructed from recycled oyster shells. Oysters are alive when both shells and valves are attached together and tightly closed when the oysters are touched or observed at low tide. If they are actively growing, the growing shell, or bill, will be thin, even translucent in sunlight, and sharp. (Be careful! Use gloves when touching or handling live oysters.) If the sill or breakwater structure is visited just after low tide, live oysters may be "spitting" water as they close their valves to protect themselves from desiccation—this is a good sign of live, healthy oysters!

Over time, oysters will die and new ones will colonize the structure. Counting dead oysters as well as live ones can give important information on the survival and self-sustainability of the developing oyster reef habitat. Dead oysters also provide refuge for small creatures and clean surfaces for new oyster spat attachment. Dead oysters (also known as "boxes") are identified by shells with both valves attached or articulated, forming an open cavity that the oyster's tissue formerly occupied.

The growth of oysters can also be monitored over time by measuring the length of the live oysters. Oyster length, or shell height, is measured from its hinge to the outermost extent of growth.[9] A small ruler (6 inches [15 cm]) can be used for this measurement. Flexible plastic rulers work best. Within each small quadrat, measure at least ten oysters of varying sizes. This can be a lot of work, so homeowners may wish to work with their team to perform this if it is required or desired.

Other organisms will be observed within the quadrats. Recording and identifying them, even if not down to the species level, will provide an indication of increases in biodiversity on living shoreline structures. Sessile organisms such as barnacles, mussels, and sponges may settle on the living shoreline infrastructure, plus mobile creatures such as crabs, snails, and shrimp should be recorded. An increase in biodiversity is a sign of a thriving habitat not just for those organisms but for the larger ecosystem. Photos should be taken of quadrats and of the species residing on the developing sill or breakwater structure. The homeowners and their teams may eventually develop an extensive photo database of the shoreline project and its biodiversity.

**FISH AND OTHER SWIMMING ORGANISMS** There are several ways to sample fish and other swimming animals in the habitats the living shoreline created. A few of the items needed can be purchased from fisheries and wildlife or environmental education supply companies. Baited minnow traps can be used to capture small fish, shrimp, or crabs that may be using installed habitats at higher tides. Beach seine nets can be pulled periodically over a set distance and location (for example, 55 to 110 yards [50 to 100 m] parallel to living shoreline breakwaters,

**FIGURE 7.18** Quadrat(s) over a section of living shoreline oysters. The numbers of live and dead oysters can be counted as well. A small subset of oysters within one or more squares of the quadrat can be measured for shell height. *Photo by Savanna Barry.*

**FIGURE 7.19** Some oyster reef-associated species. *(A)* Blue crab in an oyster structure. *(B)* Redfish. *(C)* Crown conch. *(D)* Oyster drill. *(E)* Shrimp. *(F)* Blue crab. *Photos A, B, C and F, UF/IFAS file photos by Tyler Jones. Photo D by Robert Aguilar, 2006, Smithsonian Environmental Research Center, Flickr (https://www.flickr.com/photos/serc_biodiversity/14987728363/; reproduced with cropping and repositioning under creative commons license CC BY 2.0 https://creativecommons.org/licenses/by/2.0/). Photo E by Florida Sea Grant.*

both in front [waterward], or behind [landward] of the structures). Both methods are intended to generally catch smaller-sized organisms such as baitfish, juveniles, and invertebrates like shrimp.[10]

Catches can be collected in water-filled buckets or similar containers to be identified and counted. Counting may be difficult at times when one or a few species are highly abundant. Abundance codes like those used in REEF Fish Counts can help estimate abundance. (See Reef Environmental Education Foundation, https://www.reef.org.) REEF abundance codes are as follows: Single = 1, Few = 2–10, Many = 11–100, Abundant = >100.

If those counts need to be customized, the biologist on the team can develop a standardized definition for data collection. Regardless, the homeowners or shoreline managers can at least keep count of the number of different species. Field guides should help identify most of the species. Photos can also be uploaded to a wildlife app like iNaturalist to help identify unknown species. Keep all critters in water and release them as soon as possible after identification and counting to ensure their survival.

Quarterly sampling will provide a reasonable picture of how species and abundances change over time. Please keep in mind fishing regulations. A special Florida Fish and Wildlife (FWC) permit may be needed to sample using these methods, which the biologist on the team can acquire if needed. At a minimum, a recreational fishing license should be obtained.

TERRESTRIAL ANIMALS: INSECTS AND BUTTERFLIES In the living shoreline planting zone and upland buffer, there may be an increase of insects and butterflies visiting these areas. (See "Upland Buffers" in chapter 8.) Sample the biodiversity of flying insects by using a butterfly or sweep net.[11] Like the seine net, the net should be swept through a fixed area and distance. Caught insects can then be identified, counted, then released. Again, if there are unknown species, a photo can be uploaded to an appropriate mobile app to help with identification.

## Part 5. Maintenance of Living Shorelines

Living shorelines require some regular ongoing maintenance. The first few years will require more maintenance while plants and structures are becoming established. Think of the living shoreline as an extension of the landscape, or "seascape." If the property is currently maintained by a landscape business, then the property manager or homeowner should be familiar with their regular maintenance activities, such as irrigation, mowing, trimming, weeding, fertilizing, and pest control. A few of those activities apply to a living shoreline planting area. Dead or dying plants

**FIGURE 7.20** Marsh invertebrates. *(A)* Fiddler crabs in salt marsh *(UF/IFAS file photo by Tyler Jones)*. *(B)* Mussel bed in marsh *(photo by Savanna Barry)*. *(C)* Purple marsh crab *(photo by Aileen Devlin, Virginia Sea Grant, 2020, Flickr [https://www.flickr.com/photos/virginiasea-grant/50516849283; reproduced with cropping under creative commons license CC BY-ND 2.0 https://creativecommons.org/licenses/by-nd/2.0/])*. *(D)* Marsh periwinkle *(photo by Mary Hollinger, 2002, courtesy Flickr [https://www.flickr.com/photos/51647007@N08/5114146243/])*.

should be removed and, in most cases, replaced. Control of invasive plants will be required under the living shoreline permits. Other traditional landscape maintenance work will not be necessary or even recommended at all and should be reduced drastically for the rest of the landscape. Fertilization and pest control, especially any involving the use of lawn chemicals, should never be performed near the water, and high and low marsh plants should never be mowed or cut. See chapter 9 for ideas to create a more shoreline-friendly landscape.

The native plants chosen for the living shoreline are specifically adapted to the zone where they were installed. Once they are well established, they will not need any additional water or fertilizers to survive and grow. Also, the presence of a living shoreline (and upland buffer) will help capture stormwater runoff and any pollutants from the rest of the landscape and prevent them from entering the water. Adding chemicals to these areas is not only unnecessary but also runs counter to that beneficial function the living shoreline provides. Mowing will not be necessary either, but trimming can be done on mangroves to maintain waterfront views. State law, as dictated by the 1996 Mangrove Trimming and Preservation Act, must be adhered to when considering trimming mangroves. While trimming is allowed, mangroves are an important natural feature on a living shoreline that reduce property damage from storms and coastal flooding. Despite their resilient nature, improper trimming will stress and weaken mangroves, reducing their ability to protect the shoreline.[12] Illegal trimming will reduce that ability even further and can incur significant fines from the state. The living shoreline team will be well versed in these regulations.

Adjustments to the size or shape of the living shoreline may be needed if it is eroding or experiencing more wave action than anticipated. If reef materials are moved or lost, they will need to be replaced. The goal of living shoreline reefs is to protect plantings, to stabilize and even accrete sediments behind the reefs, and therefore, to widen the shoreline. It is possible that sediment deposition will be significant enough to start burying the reefs, reducing the available oyster habitat. As shorelines are dynamic, there may be periods when reefs become buried and uncovered. Monitoring over time will increase the homeowners' or shoreline managers' understanding of these cycles. But if the reefs are becoming more consistently buried, adding reef materials to regain their height and optimal tidal elevation may be recommended. The team will help determine this and whether existing permits will allow it, or if permit modifications will be needed.

The importance of maintaining a living shoreline property cannot be overstated. An established living shoreline is low maintenance but not a one-time "set it and forget it" endeavor. A functional living shoreline will be an evolving

**FIGURE 7.21** Mangrove trees can be trimmed to preserve a view, depending on their height. (See also height ranges for legal mangrove trimming figure 7.22.) *(1)* Hedge-trimmed mangroves can be combined with *(2)* "windowed" mangroves to provide some shoreline protection while satisfying aesthetic needs. *Illustration by Marjorie Shropshire.*

part of the waterfront. Proper and consistent ongoing maintenance will ensure its long-term functioning and aesthetic. The team can help to develop a long-term maintenance plan.

## What to Expect

Living shorelines will improve over time as plants become lush and fill in; and as living oysters colonize the reef, more wildlife, birds, and fish will find habitat there. There should not be a lot to do, other than regular maintenance like removing

### Mangrove Trimming Guidelines

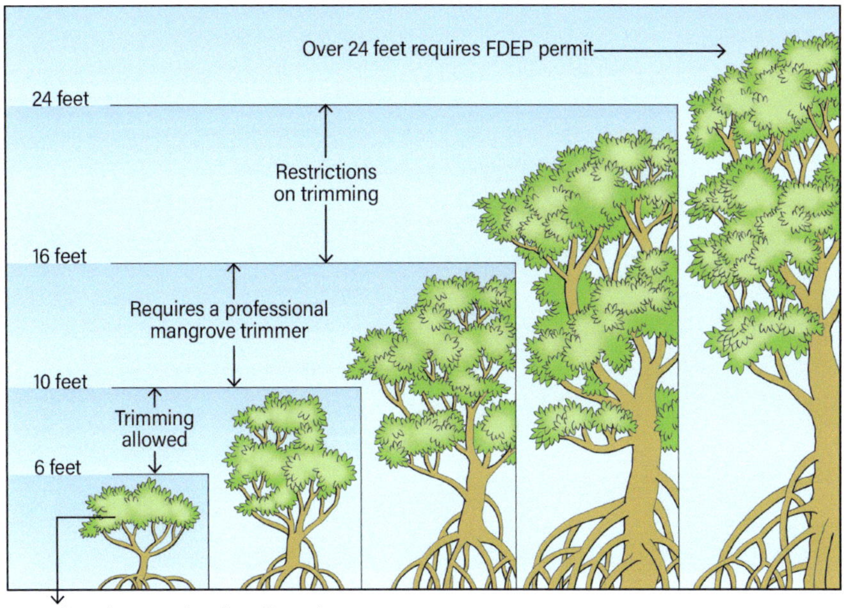


**FIGURE 7.22** Height ranges for legal mangrove trimming. Mangroves under 6 feet tall (1.8 m) cannot be trimmed. Mangroves over 24 feet tall (7.3 m) require a permit from the Florida Department of Environmental Protection (FDEP). Mangroves between 6 feet and 24 feet (1.8 m to 7.3 m) have varying requirements. *Illustration by Marjorie Shropshire. Graphic adapted from FDEP, "Trimming Mangroves," February 8, 2016, https://floridadep.gov/sites/default/files/mangrove-trimming-2-08-16.pdf.*

#### THE 1996 MANGROVE TRIMMING AND PRESERVATION ACT

See the Florida Department of Environmental Protection, "Mangrove Trimming and Preservation Act" (MTPA) (https://floridadep.gov/water/submerged-lands-environmental-resources-coordination/documents/mangrove-trimming-and) and their link to the *Mangrove Trimming Guidelines for Homeowners* (https://floridadep.gov/sites/default/files/Mangrove-Homeowner-Guide-sm_0.pdf). Also be aware that local governments are delegated authorities for the 1996 MTPA. Delegated authorities may have regulations in addition to those of the state that a waterfront property owner and living shoreline professional should be aware of in their specific locations.

**FIGURE 7.23** A beautiful example of a mature living shoreline in a residential setting. For preliminary sketches and before photo, see figure on page 164. *Photo by Ocean Research and Conservation Association Inc. (ORCA)*

invasive plants and perhaps replacing a damaged plant on occasion. A living shoreline will provide unique access to the water and its nature that homeowners and their families can enjoy for many years.

## Conclusion

Building a living shoreline is a dynamic process that requires careful planning, thoughtful design, and dedicated teamwork. From preparing the site to installing natural elements, each step contributes to the overall success of the shoreline. Whether homeowners or shoreline managers choose a nature-based approach or a hybrid solution, a living shoreline will provide numerous benefits for the environment and will become part of the local shoreline ecosystem. A successful living shoreline will also stabilize the shoreline and increase its resiliency to imminent and future threats as climate change causes increased storm impacts and rising seas.

## Notes


1. Justin T. Ridge, Antonio B. Rodriguez, F. Joel Fodrie, et al. "Maximizing Oyster-Reef Growth Supports Green Infrastructure with Accelerating Sea-Level Rise," *Scientific Reports* 5 (2015): 14785, https://www.nature.com/articles/srep14785.
2. Mollie F. Duggan-Edwards, Jordi F. Pagès, Stuart R. Jenkins, Tjeerd J. Bouma, and Martin W. Skov, "External Conditions Drive Optimal Planting Configurations for Salt Marsh Restoration," *Journal of Applied Ecology* 57 (2020): 619–29; Hallie S. Fischman, Sinead M. Crotty, and Christine Angelini, "Optimizing Coastal Restoration with the Stress Gradient Hypothesis," *Proceedings of the Royal Society B* 286 (2019): 20191978, http://doi.org/10.1098/rspb.2019.1978.
3. Eric L. Sparks, Just Cebrian, Patrick D. Biber, Kate L. Sheehan, and Craig R. Tobias, "Cost-Effectiveness of Two Small-Scale Salt Marsh Restoration Designs," *Ecological Engineering* 53 (2013): 250–56.
4. Brian R. Silliman, Elizabeth Schrack, Qiang He, et al., "Facilitation Shifts Paradigms and Can Amplify Coastal Restoration Efforts," *Proceedings of the National Academy of Sciences* 112, no. 46 (2015): 14295–300, https://www.pnas.org/doi/10.1073/pnas.1515297112.
5. Laura Reynolds, Natalie C. Stephens, Savanna C. Barry, and Ashley R. Smyth, "Living Shoreline Monitoring—How Do I Evaluate the Environmental Benefits of My Living Shoreline? SS694/SL481, 1/2021," *EDIS* 2021, no. (2021), https://doi.org/10.32473/edis-ss694-2021.
6. Reynolds et al., "Living Shoreline Monitoring."

7. David Cannon, Kelly Kibler, Melinda Donnelly, et al. "Hydrodynamic Habitat Thresholds for Mangrove Vegetation on the Shorelines of a Microtidal Estuarine Lagoon," *Ecological Engineering* 158 (2020): 106070, https://www.sciencedirect.com/science/article/pii/S092585742030358X.
8. J. B. Kauffman and D. C. Donato, "Protocols for the Measurement, Monitoring, and Reporting of Structure, Biomass, and Carbon Stocks in Mangrove Forests," Working Paper (Center for International Forestry Research, 2012), 10–20, https://www.cifor-icraf.org/knowledge/publication/3749/.
9. Lesley Baggett, Sean Powers, Robert Brumbaugh, et al. *Oyster Habitat Restoration: Monitoring and Assessment Handbook* (The Nature Conservancy, Arlington, VA, 2014), https://www.conservationgateway.org/ConservationPractices/Marine/Pages/oystermonitoring-handbook.aspx.
10. Laura Reynolds, et al., "Living Shoreline Monitoring."
11. Reynolds et al., "Living Shoreline Monitoring."
12. Natalia Medina Irizarry and Michael Andreu, "The Impacts of Trimming Mangroves: FOR378/FR448, 02 2022," *EDIS* 1 (2022): https://doi.org/10.32473/edis-fr448-2022.

# 8

# Building Buffered Shorelines and Upland Buffers

## Protecting Shorelines with Buffers

In urban or suburban coastal neighborhoods, shorelines can be protected and improved with plant-based buffers, which are lightly maintained non-lawn areas between a shoreline or wetland and the rest of the landscape. Buffers can be wooded with a diversity of native trees, shrubs, and understory plants; herbaceous with bunching grasses, rushes, sedges, vines, and wildflowers; or a combination to suit the property and how it is used. Buffers protect the shoreline properties from erosion by both wind and water, and they help to improve the water quality by absorbing stormwater runoff, nutrients, and other materials from the upland landscapes. Buffers may also provide habitat for shoreline birds and other wildlife.

Building buffered areas upland from living shorelines is strongly recommended to enhance and protect the living shoreline plants and structures. The selection of plants and the methods and arrangements of the plantings will vary depending upon the nature of the shoreline property.

Buffers are also highly recommended for shoreline properties where living shorelines are not feasible, such as high-energy shorelines, canal-front shorelines where there is limited space between the landscape structures and the mean high water line (MHW), or properties abutting wetlands.

Even if a property abuts a wetland or body of water, permits are not usually required to install or enhance buffers above the mean high water (MHW) line. However, a permit may be required, if the property is part of, or adjacent to, an easement or conservation area, or if the plans include modifying the size of or directing additional stormwater into existing natural ponds (including seasonal ponds) or artificial retention ponds. Check with the local authorities to verify the permit situation before installing any plants.

Buffers should be designed to do the following:

- Absorb stormwater and surface runoff and any nutrients or pollutants it carries from the upland landscape
- Trap or reduce air-borne soil, dust, sand, or other materials
- Trap deposits of grass clippings, litter, and other organic and inorganic materials that may be generated by maintenance and other activities on upland suburban and urban properties

**FIGURE 8.1** Grass clippings floating on the water add to the nutrient-rich scum. Replacing lawns next to waterside areas with buffer plantings will reduce this source of water pollution. *Photo by Ginny Stibolt.*

For this chapter, the buffered areas near shorelines are divided into two sections: high marsh buffers and upland buffers.

(1) **High Marsh Buffers** are installed between the mean high water (MHW) line and the mean higher-high water (MHHW) line. These areas may experience higher tides several times a year during king tides or storms. (See chapter 1 for more information on tides.)

(2) **Upland Buffers** are installed upslope from the high marsh area. Depending on the location and the structure of the shoreline, this area of the landscape may experience salt spray and occasional flooding during storm surges, so upland buffers will need to be designed to withstand these events.

If regular access to the shoreline or waterway through the buffered area is needed, the access should not impede the functioning of the buffer to absorb runoff from the landscape or to stabilize the shoreline soils. The access pathways could be constructed with elevated boardwalks or made with large, heavy steppingstones that will not wash away during a storm surge. If the access pathway is on the ground, it is a good idea to build it with curves or slant it across the slope to reduce the possibility of erosion from the landscape.

Building or expanding buffered areas is an important undertaking to develop a more shoreline-friendly landscape. There are several ways to get started, and the most efficient method will depend upon the nature of the property and how it is used. In most cases, buffers installed on waterfront common areas in neighborhoods or municipalities will require different approaches and designs than privately owned shorelines.

## High Marsh

Depending upon the property characteristics above living shorelines, high marsh plantings are usually installed after all the initial work on the living shoreline is completed, possibly even after the living shoreline plants have become established. Since this section of the landscape will be under water only a few times a year, a higher diversity of more upland plants can be used. Contractors could install the buffer plantings in this area as part of the living shoreline project, but homeowners, shoreline managers, or volunteers could also install these plantings, since this part of the shoreline is drier and more accessible than the living shoreline installation site. In contrast to the living shoreline, which is exposed to tidal waters once or twice a day, the shoreline buffer plantings will need frequent irrigation until the plants are established, and they will need regular irrigation during droughts for a year after that.

**FIGURE 8.2** An example of a high-energy site that is not appropriate for a living shoreline, but where a shoreline buffer could be installed adjacent to the seawall and dock to increase resilience. Evidence of wave or tidal action is indicated by the red mangrove propagules that have washed over the seawall and accumulated on the dock and shore. *Photo by Marjorie Shropshire.*

High water events now over-top or breach many of the older bulkheads in Florida due to sea level rise, king tides, and storms. This means that buffered shoreline plantings are now needed on the upland side of these regularly flooded bulkheads to protect the shoreline, wetlands, and the water.

In most cases, high marsh plantings, whether installed above living shorelines or above older bulkheads, should be designed with bioretention swales located on the upland side of the buffer plantings to most effectively absorb the stormwater runoff from the rest of the landscape.

In addition, planting salt-tolerant trees and shrubs with significant leaf surface area in groupings, or groves, in these high marsh areas will soak up more water through transpiration than smaller herbaceous plants. If there are single trees already located in the high marsh, it is a good idea to build a grove of compatible trees and shrubs around it not only to soak up more water but also because a grove of woody plants is more wind tolerant than a single tree and provides a better environment for tree growth and health than being surrounded by lawn.

## HOW PLANTS WORK

### TRANSPIRATION

For terrestrial ecosystems, the transpiration process of vascular plants is a major influence in ecosystems because it moves large amounts of water from the soil to the air. This water movement is an important process that cools the air and produces rain in the area.

**Transpiration: How Water Moves Through a Plant**

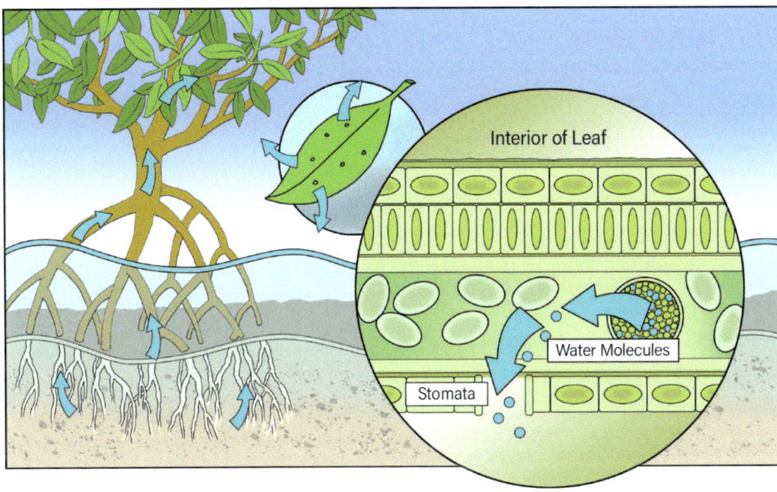


**FIGURE 8.3** Moisture in the soil is absorbed into the roots through osmosis and is transported through tiny tubes formed by the xylem cells through the plant. About 90 percent of this water evaporates into the air through the leaf pores (stomata). *Illustration by Ginny Stibolt and Marjorie Shropshire.*

Vascular plants absorb water through their roots through osmosis. The water then moves upward through the plants through their xylem cells, which form tubes that run throughout the plant. The water moves into the leaves where it evaporates into the air through the stomata (pores) on the leaf surfaces. More than 90 percent of this absorbed water runs right through a plant. The remaining 10 percent of the water is used for photosynthesis and other metabolic processes, and to keep the leaves and stems turgid so they do not wilt.

Since water adheres to itself as it is drawn through the plant's xylem tubes, it pulls more water with it as it moves up. The big driver is the difference between concentrations of the water in the soil, the plant, and the air. Nature wants everything to be equal, so this differential is what sucks the water up through the plants.

Most plants close their stomata at night or when a plant wilts. The guard cells on either side of the stomata close to preserve water in the plant, and when this happens, transpiration stops and usually photosynthesis as well. Since most transpiration takes place during daylight, this is also when roots are actively absorbing water from the soil, so the best time to irrigate plants is early in the morning, just as transpiration is getting started for the day.

Also, because of water evaporating from plants, the temperature is lowered. It varies depending on the ambient temperature, but "using the sun's energy, individual trees can transpire hundreds of liters of water per day. This represents a cooling power equivalent to 70 kWh for every 100 L of water transpired (enough to power two average household central air-conditioning units per day). With deeper roots, trees can maintain their cooling function even during long-lasting heatwaves."[a]

This cooling effect reduces the latent heat in the immediate area of a tree. In other words, the physical process of water shifting from a liquid inside the plant to a vapor in the air during transpiration causes the temperatures to be lowered. It is cool in a forest not only because of the shade but also because the air temperature is actually lower. This is also what happens when we sweat—our bodies are cooled due to the evaporation of water.

The plants with the most leaf surface area have the most stomata and therefore transpire the most water. When designing the plantings for a bioretention area or a rain garden, use broad-leafed trees and shrubs when possible. A mature oak tree can transpire more than 400 liters of water on a hot summer's day. In Florida, this is an important consideration.

[a] David Ellison et al. "Trees, Forests and Water: Cool Insights for a Hot World," *Global Environmental Change* 43 (2017): 51–61, https://www.sciencedirect.com/science/article/pii/S0959378017300134?via%3Dihub#bib0200.

## Upland Buffers

Since the upland buffer areas are rarely subjected to storm surges, the installation of upland buffer plants is more like other landscaping projects, but it still has the overarching mission of protecting the shoreline and the wetlands or body of water beyond it. Also, upland buffer plantings need to be designed for wind tolerance, salt-spray tolerance, and water absorption. When upland buffer areas replace lawns, it improves the water quality right away, especially if the lawns had been maintained with pesticides and synthetic fertilizers. (See the discussion on lawns in chapter 9.)

## A MUNICIPAL INCENTIVE

As of this writing, Satellite Beach, Florida, provides monetary incentives to encourage homeowners and other shoreline managers to install buffers next to shorelines or bulkheads to improve water quality. Scientists working for the city tested the water quality before and after the buffer installations to demonstrate that buffers reduce the pollutants in the water. This program and its minimum requirements can be used as a general guideline for effective buffers or as an example for other communities that want to use this citizen-driven method to improve their water quality.

Here are the minimums set for buffers in order to qualify for matching funding:

### VEGETATED BUFFER MINIMUM PARAMETERS

- Minimum of 5 ft. (1.5m) in depth from seawall or canal edge
- Minimum of 20 ft. (6.1m) in length
- Plants within vegetative buffer will consist of native flora only
- Vegetative buffer shall contain a minimum of 3 plant species from suggested plant list
- Buffer zone shall contain inorganic substrate (e.g., sand, crushed coquina, shell, rock), edging will be required per code for washout prevention (for inorganic substrate)
- Swale is suggested for runoff and sediment retention
- Work with City on design and allow access once established for photo[a]

[a] City of Satellite Beach, "Vegetative Buffer Zone Plant Grant," accessed January 31, 2025, https://satellitebeach.gov/residents/sustainable_satellite/vegetated_buffer.php.

## Converting Lawn to Buffer

If lawn extends to the shoreline, the edge of a wetland, or to the edge of a bulkhead, there are various methods to replace it with a functioning buffer area. However, there are a few decisions to make before getting started, including:

- Whether to install the whole buffer area in one season or to build it in several stages over a few years as a multi-stage project.
- Where to place the bioswales. (They absorb the stormwater runoff from the property and greatly improve the effectiveness of the buffer.)

**FIGURE 8.4** Two public spaces designed to absorb water from higher high tides and stormwater. *(A)* A dry detention marsh at the Ripple Stormwater Eco-Art Project in Old Palm City, one of four sites planted with native vegetation that prevent neighborhood stormwater from entering the nearby St. Lucie River. *(B)* Tidal Garden in Palm Beach was installed in a public park where the high-water events over-top the seawall several times a year. The garden was designed to absorb and contain those higher tides, to contain stormwater, and to provide public access and infrastructure such as steppingstones and stone benches. *Photos by Ginny Stibolt.*

Some homeowners or shoreline managers will hire contractors or landscapers to remove the lawn, construct the bioswales, and install all the buffer plants within a few weeks; others may take it slower and install the buffer areas in stages. There are pros and cons to each approach. On the one hand, if the buffer is built over several years, homeowners and volunteers may be able to build it without outside help to save money. In addition, the trees could be planted first to become well established before the understory trees and shrubs are added a few years later. On the other hand, a big advantage of creating the upland buffer quickly is that work of establishing all the plants, including extra irrigation, is concentrated into one or two seasons, and the plants will begin to function as a buffer more quickly and effectively than if the project stretches out over several years.

Whether the work is done all in one season or over the course of a few years, the installation of a temporary drip irrigation system is recommended so that all the new plants receive enough irrigation to develop good root systems in their new locations. After two or three years, or once all the native plants are established, the irrigation system can be removed, because in most cases the native plantings will not need supplemental irrigation after establishment.

## Lawn Removal

In most cases, it is best if the lawn is removed or killed without poisons, but in some cases, using an herbicide may be the most effective method, especially if there are deep-rooted, nonnative weeds such as torpedo grass (*Panicum repens*) or Florida snow (*Richardia* spp.), which thrive in damp spots and often stretch out into the water. Use an herbicide that is approved for wetland or wetland-adjacent use, and be sure to closely follow the label instructions. Apply the herbicide only when the air is still and when there's little chance of rain for the next few days to ensure that herbicide does not reach the wetland.

### LAWN REMOVAL WITHOUT POISONS

- Physical removal of lawn plants using hand or power tools is the fastest way to accomplish lawn removal, but this method includes a lot of physical labor, even with a professional-grade sod cutter. The lawn plants must be completely removed from the site, or they could take root again.
- Solarization (where the area is covered with clear plastic and left in place for several weeks) kills the grass and other plants with roots near the soil surface. This method is most effective in the summer, but it may not kill plants with deep roots, tubers, or rhizomes. If there is an irrigation system in place, solarization heats up the whole area and this

could damage plastic parts on sprayer heads that are near the surface. It is possible, however, to temporarily remove sprayer heads so they will not be damaged.

- Cover the lawn with something that will not blow away and is thick enough so no light can reach the plants. As part of this option, it is best if there is a complete separation from the rest of the lawn, so the part that is buried is not supported by crawling lawn plants such as St. Augustine grass that are still partly uncovered. Building a bioswale on the upland side of the buffer will usually take care of this problem.

## Planting Buffers

Once the lawn has been removed and the waiting time is over, if herbicides were used, then installing the plants can begin. If no plants will be grown from seed in the newly cleared area, corn gluten or some other germination inhibitor can be applied in the area to reduce the germination of weed seeds. It will eventually turn into an organic nutrient. Pine needle or wood chip mulches also act as germination inhibitors in more upland areas.

## Mulching Guidelines

For a shoreline-friendly property, there should be no chance of mulch ever ending up in the adjacent waterway or wetland. Many people recommend that areas of the landscape below the MHHW line, which is reached by king tides and storm-induced high water events several times a year, should be mulched with inorganic products such as coarse sand, gravel, or crushed shell that will not release nutrients into the water. However, these mulches do not moderate the soil temperatures and do little to reduce weed seed germination, so they are best used closer to the water in high marsh areas, while organic mulches are best for more upland sites.

Organic mulches, such as wood chips, pine needles, or straw, reduce soil erosion, reduce weeds, reduce moisture loss, and moderate temperature changes in the soil. In addition, as they decompose, the soil under the mulch will develop a temporary nitrogen deficit as the decomposers metabolize the material, which kills weed seeds in that layer. Eventually, these mulches add to the soil layers. Wood chips or chopped bark float and for this reason they should not be used in landscapes below the MHHW line, because the water from the king tides or storms can wash them away. (See appendix B for more details and the pros and cons of various types of mulching materials for shoreline properties.)

## Mulch for New Plantings

Mulching is part of the process in developing all new planting areas such as pollinator gardens, meadow areas, rain gardens, or groves of trees and shrubs, not just buffer areas. After installing the plants, lay in the mulch about 1 to 2 inches (2.5 to 5.0 cm) thick in between the plants—but not touching the stems or trunks. For larger woody plants that may need extended irrigation, build a berm of soil just outside of the planting hole to retain the extra water, but cover the area with mulch to help keep the moisture in the soil. (See chapter 9 for more details on planting trees.)

While it is best to buy young trees and shrubs so they have the best chance of survival, they should be planted so there is enough room for their mature sizes. This will leave big spaces between the plants; so while waiting for these plants to mature, filling in the spaces between them with a thick layer of mulch such as pine needles, which is unlikely to blow away or float away and will not form a water-resistant crust, is recommended. Also, plant fast-growing ground covers such as dune sunflowers (*Helianthus* spp.), fog fruit (*Phyla nodiflora*), railroad vine (*Ipomoea pes-caprae*), bunching grasses, or other site-appropriate meadow plants to discourage weeds to some extent and to hold the soil in place. There will still be some manual weed control needed, especially the first two or three years as the buffer plants mature.

In a few years, as the trees and shrubs mature and begin to cast significant shade, some of these initial sun-loving filler plants may need to be transplanted to sunnier locations and replaced with ferns or other more shade-tolerant plants. The initial planting and mulching between the trees and shrubs simulate a natural leaf mulch to keep the soil cool, to reduce weed germination, and to reduce erosion until the plants can mostly do this for themselves.

## Mulch for Paths

Mulches used for covering unpaved paths or between steppingstones will need to be thicker—maybe 6 inches (15 cm) deep. In this case, as the mulch becomes part of the soil, new layers of mulch will be needed every year or two depending on the traffic and the type of mulch. The goal here is to keep the soil covered to protect it from foot or vehicular traffic and to prevent erosion by wind or water.

**FIGURE 8.5** Pine needles are a good mulch for transition areas between mowed areas and wilder areas or wooded areas. Shown here is muhly grass (*Muhlenbergia capillaris*) that is planted in a transition area between a wooded area and a mowed area. *Photo by Ginny Stibolt.*

## Mulch for Edging

Mulches help to maintain defined mowing edges along the transition areas between mowed areas and other habitats. The goal is for the mower to make one sweep along the edge with no backtracking and no additional trimming. Depending upon the situation, these mulches could be renewed every two or three years. To slow down the growth of lawn grasses into the transition areas, fold or bend back the stems or rhizomes of the crawling grasses and weave them away from the edge instead of cutting them off at the edge. (The cutting will stimulate the grass to grow two or more new shoots just below the cut, which will create even more work.) Also make sure that the mulch will not be lofted by the mower by tamping the mulch down along the edges of the mowed areas.

## Layering Under the Mulch

For large mulching projects, many people use layers of cardboard (with all the staples and tape removed) or layers of newspapers on the soil and then cover the paper with a thick layer of organic mulch like wood chips or pine needles. This quickly kills most plants growing under the layers, so this method is often used to remove lawn areas. The paper will eventually decompose, and the grass underneath will also become part of the soil, but make sure the mulch is thick enough, so it does not blow off or rinse off in a storm. Cut through the paper layer to install plants (large or small) in the area so their roots have access to the soil below.

Geotextile or weed barrier cloth is not recommended for reducing weeds in Florida landscapes. Weeds grow on top of it and under it, and roots of nearby plants will soon claim it as their own, which will increase the maintenance chores and disrupt the landscape when it is finally removed. Weed barrier cloth also reduces the habitat values for many important soil inhabitants such as worms, toads, and native bees. In addition, this synthetic fabric also contributes to plastic pollution in the landscape. However, geotextiles are useful in the construction of drywells to keep soil from intruding into the catchment space. (See the drywell photos in chapter 9.)

## Bioswales in Buffers

Most shoreline landscapes slope to some extent toward the shoreline or wetland. To effectively prevent stormwater runoff from reaching the wetland or body of water, the upland buffer area should include at least one bioswale that runs the entire length of the shoreline. If the slope toward the shoreline is relatively uniform, then the bioswale should be parallel to the shoreline. If there is a steeper slope in one area of the landscape, then the buffer should be wider for that section, and contoured to fit the slope. The placement of the bioswale(s) should be decided and installed before buffer planting begins.

The most important bioswale location is on the upland side of the buffer to capture stormwater runoff from the landscape. Additional bioswales can be closer to the shoreline, between the buffer plantings or just upland of the shoreline. Bioswales should be planted with rain garden plants to absorb stormwater and hold the soil. (See chapter 9 for more information on rain gardens.)

It is best if the upland section of the landscape is densely planted with trees, shrubs, bunching grasses, ground covers, permeable paving, or other landscape features that are designed to sequester stormwater and hold soil in place. This way, eroded soil will not fill in the bioswales.

**FIGURE 8.6** Weed barrier cloth or geotextile does not block weeds in Florida, since weed roots will penetrate it from above and below, which makes managing weeds more difficult. In addition, the barrier formed by the geotextile prevents movement of important soil inhabitants such as worms, toads, and native bees. *Note:* The use of gravel as mulch also does not deter weeds. *Photo by Ginny Stibolt.*

## Enhancing an Existing Buffer

Planting trees and shrubs in groups is more effective for wind resistance than individual trees planted alone. If a solo tree exists in the planned buffer zone, then it is a good idea to plant other trees and shrubs around it to create a "grove" and to simulate the way plants grow in nature. Consideration can be given when locating groves to preserve a desired view.

If the property already has a wild or semiwild area between the shoreline and the maintained landscape, such as a lawn or garden areas, there are various ways to enhance its effectiveness as a buffer.

If there are mature trees directly abutting a lawn or mowed area, it is a good idea to create a transition area by planting groupings of shrubs and understory trees on the upland side of the wooded area. Install a bioswale along the whole area, but at least 10 feet (3.1 m) away from the trees so that building it will not disturb the surface roots of trees and shrubs. The bioswale should be designed to trap any runoff

**Bioswale in an Upland Buffer**

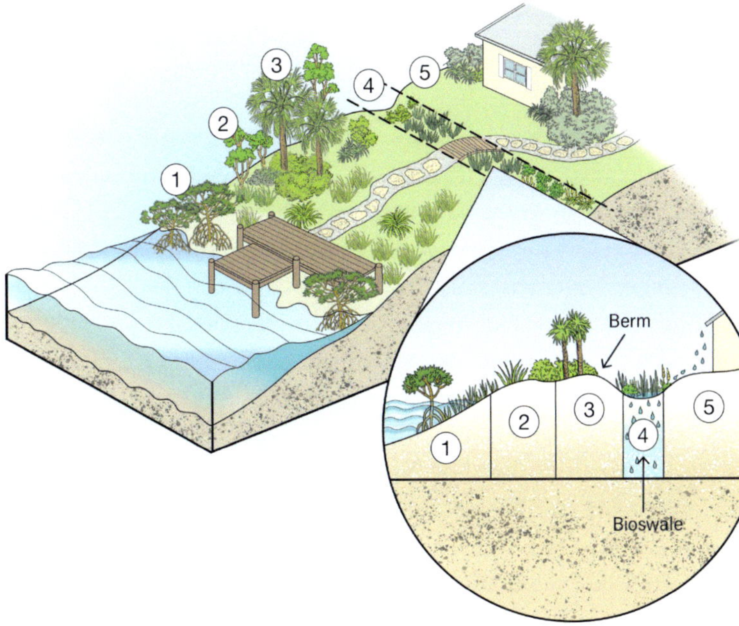


**FIGURE 8.7** After a rainstorm, stormwater runs down slope from buildings, driveways and other impermeable infrastructure *(5)*. A bioswale *(4)* captures stormwater runoff before it reaches the upland buffer *(3)*. Vegetation in the buffer area *(2, 3)* absorbs stormwater from the bioswale and soil before it enters the intertidal zone *(1)*. *Illustration by Marjorie Shropshire.*

**Enhancing an Existing Buffer**

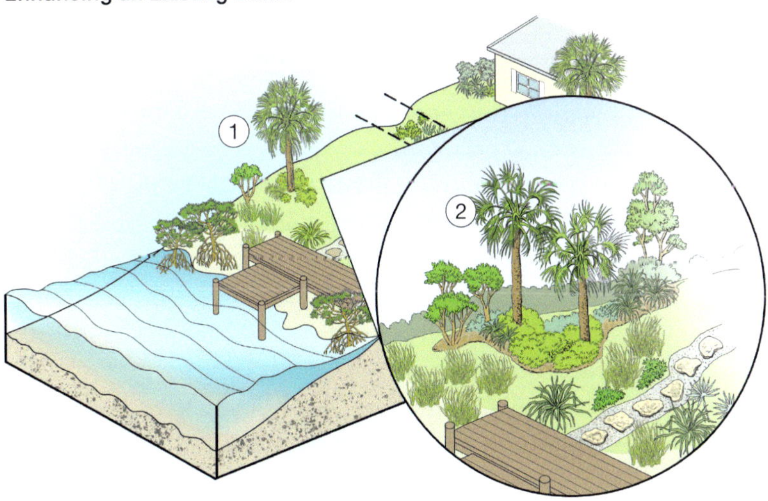


**FIGURE 8.8** The function of an existing buffer can be enhanced by adding additional plants. Add plants around *(1)* solitary trees and shrubs to create *(2)* groups or groves of plants for more effective wind resistance. *Illustration by Marjorie Shropshire.*

from the rest of the landscape, and it can be incorporated as part of this new transition area and covered with tough ground cover plants and rain garden plants.

If the non-lawn area has only herbaceous plants, and if it is not a damp area with frequent inundations like a salt marsh or a swamp, then it may be a good idea to build one or more groves of assorted trees and shrubs to provide a rounded, wind-tolerant arrangement of woody plants. The trees and shrubs provide better shoreline stability, better absorption of water, and some protection from storm surges and onshore winds. As a bonus, these groves will cool the air and provide habitat for birds. A bioswale on the upland side of this area is recommended to provide more buffering from stormwater runoff and from other materials that might wash toward the shoreline.

It is also important to design the buffer plantings so that there is access to all parts of it for ongoing maintenance, especially for irrigating, pruning, mulching, and weeding (especially the removal of known invasive plants). These pathways should be narrow and would usually be parallel to the shoreline, so they do not provide easy runoff locations. Ongoing maintenance will be greatly reduced over the years, but it will not disappear entirely.

## Plant Recommendations

Choose plants using either of the two lists provided in appendix A, depending upon the conditions of the buffer's area. The first list includes plants that are at least somewhat tolerant of salt spray or inundation with brackish or saltwater. The second list includes non-salt-tolerant plants that would be more appropriate for environments with freshwater or nearly freshwater.

The Florida Native Plant Society (https://www.fnps.org/plants) has an online tool to help users select recommended native plants that are native to a specific county. This tool also includes criteria on whether plants are salt tolerant or whether they are recommended for rain gardens. Once a list of plants has been created, select the plant profile for each of the plants, which includes information on moisture and light requirements for that specific plant and provides a link to native plant vendors that have it in stock.

The Florida Association of Native Nurseries (FANN; http://www.plantreal-florida.org) has a website where users can locate native plant vendors who have a specific plant or find local native plant vendors. Many of the members also offer landscaping services in addition to having good local knowledge of native plants.

Information on best practices for installing buffer plants and for information on caring for the plants until they are well established is covered in chapter 9.

## Ongoing Maintenance of Buffers

Even though building buffer areas by installing native plants in natural-looking groupings could be considered a type of restoration, and even though buffers may be designed to look like a natural area, they are not wild and will need ongoing maintenance. That said, one important goal of designing a native plant buffer is to have it become more self-sustaining as the plants mature and as it matures into a working ecosystem.

Over the first few years, remulching of areas between the plants will be needed to keep the soil covered to reduce erosion by water or wind, to reduce the weeds, to moderate the soil temperature, and to eventually enrich the soil. Make sure that the mulch does not touch the trunks of the trees or shrubs and that no mulch ends up in the adjacent wetland or waterway. As the plants mature, they will produce more of their own mulch consisting of fallen leaves and small branches, so the need for outside mulching will be reduced over the years, except perhaps for the upland side of the buffer where it intersects with the lawn or other landscape features. Keeping a defined mowing edge with mulch will probably be an ongoing project that will need to be repeated every few years.

Even with a good layer of mulch on the soil and with dense plantings in the buffer, in Florida, new plants will invite themselves to the disturbed soil. These are called pioneer plants, and their seeds are everywhere and will sprout wherever there is disturbed soil. The dominant pioneer plants vary depending upon location and climate. In some cases, the pioneer plants may be natives that will enhance the buffer's ecosystem, but oftentimes those new plants may be too aggressive for the new plantings. Unfortunately, many aggressive plants are known invasive plants, as determined by the Florida Invasive Species Council. Such invasive plants have taken over wild spaces in the state and will need to be removed. (See a discussion on invasive plants in chapter 2.)

It is a good idea to bring in an expert to help identify the pioneer plants in the new buffer to help make the determination of whether they should be pulled out or left to become part of the buffer. Help with this could come from the living shorelines team for that property or from the local cooperative extension office. (See the University of Florida's IFAS Extension page "Find Your Local Extension Office by Region," https://sfyl.ifas.ufl.edu/find-your-local-office/.)

It is best to pull or dig these weedy plants rather than to use any type of herbicide—commercial or homemade—near a shoreline and near the new buffer plants. Once the weeds are recognizable and identified, pull them while they are still small. Pulling the weeds while they are small, tamping down the soil, and

**FIGURE 8.9** A shoreline buffer of diverse and attractive native plants has been installed landward of a trimmed mangrove fringe at this waterfront community in Martin County, Florida. The buffer serves to keep contaminants from the lawn area out of the nearby waterway. *Photo by Cameron Maletto, Longleaf Design LLC.*

adding mulch several times for the first couple of years will disrupt the soil less than if the weeds mature and go to seed. Disturbing the soil as little as possible while weeding will reduce future weed populations, because every time the soil is disturbed, new pioneer plants can get a start.

After the trees and shrubs have grown enough to cast some significant shade and the soil surface is no longer disturbed regularly, the uninvited plants will be reduced to some extent. While the known invasive plants will always need to be removed, as the years go by, the aggressive native weeds are likely to be less of a problem.

After trees and shrubs are established in the buffer zone, very little pruning should be needed unless branches are broken during storms. In fact, in a buffer, the goal is for the plants to form a wind-tolerant rounded mass. If a tree dies, leave it in place as a snag, if it will not cause damage to property when it falls, because the snag will add wildlife habitat value. (See chapter 9 for initial pruning guidelines.)

## Buffering

Building buffered areas is an investment of time and money, and it will take years for the woody plants to mature; but their installation along with the bunching grasses and other herbaceous plants and groundcovers will help to improve the water quality right away.

## Resources

Environmental Law Institute, *Planner's Guide to Wetland Buffers for Local Governments* (March 2008), https://www.ecosystemmarketplace.com/wp-content/uploads/archive/documents/Doc_457.pdf.

Gary Bentrup, *Conservation Buffers: Design Guidelines for Buffers, Corridors, and Greenways* (US Department of Agriculture and US Forest Service, 2008), https://www.fs.usda.gov/nac/buffers/docs/conservation_buffers.pdf.

Homeowners' Guide to Wetlands and Buffers. https://sarasota.wateratlas.usf.edu/upload/documents/Homeowners-guide-to-wetlands-web.pdf

# 9

# Shoreline-Friendly Landscaping

**WHILE THIS BOOK'S MAIN** topic is living shorelines, there are several strategies, in addition to planting shoreline buffers, that can make any property more "shoreline friendly," whether it abuts a shoreline or not. These actions will improve water quality and improve aquatic and wetland habitats and can, with some exceptions, be accomplished by homeowners, shoreline managers, or volunteers and without the need for permits or outside contractors.

Shoreline-friendly actions include the following:

- Stopping all landscape-wide applications of pesticides and fertilizers
- Reducing landscape-wide irrigation
- Stabilizing the soil to reduce erosion
- Retaining most of the stormwater
- Making sure that septic systems are working correctly
- Replacing solid surfaces with those that are more permeable
- Controlling pets and their waste
- Planting more trees to hold soil, increase water uptake, sequester more carbon, and cool the air

All these actions protect wetlands or bodies of water including estuaries, rivers, ponds, lakes, and freshwater or saltwater marshes by reducing, diverting, and absorbing stormwater runoff that is often loaded with eroded soil, fertilizers, pesticides, and other pollutants.

The most effective actions to make a property more shoreline friendly will vary depending upon the property's size, the main uses of property, and the agenda of the owners or managers of the property. There are various components, as described below, that can be implemented in any order and that may be implemented in a short amount of time with contractors and experts handling most of the details; or they can be implemented one step at a time over a few years. Either way, any of these actions will enhance the shoreline ecosystems and improve the water quality.

## Rethinking the Lawn

Rethinking the lawn is the simplest and most consequential action a homeowner or property manager can take to make a landscape more shoreline friendly. There are four million acres of lawn in Florida,[1] and historically many homeowners and property managers have maintained lawns by applying:

- herbicides to kill non-grass plants,
- fungicides to reduce fungal diseases,
- insecticides to kill the insects that might consume the leaves or roots of the turfgrass, and
- synthetic fertilizers to offset the ill effects of the other applications.

Lawn-care companies, homeowners' associations, and other organizations may have also recommended more frequent irrigation, overseeding with winter grasses, and other measures to keep the turf growing and green all the time.

However, a recent study documented the following:

> *There are environmental concerns associated with turfgrass, including chemical runoff from fertilizers and pesticides which can affect the environment.*

Turfgrass fertilizer has been connected to non-point-source water pollution. This has led to algal blooms and depleted oxygen levels in waterways and nutrient enrichment in many water bodies in Florida. The study also found that fertilizer type and its application (amount/frequency), irrigation, and rain frequency affect the level of nutrient leaching from residential lawns. Furthermore, the state of Florida has put into place regulations that control fertilizer use on residential turfgrass to reduce non-point-source pollution.[2]

This same study found that once Florida homeowners were educated about this issue, they are more likely to favor "low-input or low-requirement" lawn maintenance.

There is a significant opportunity for homeowners, municipalities, homeowners' associations, and other shoreline property managers to make a real difference in the local water quality while spending less money on lawn care. There are also significant opportunities for the landscape industry to play a positive role by acting as contractors for various shoreline managers in designing, installing, and maintaining living shorelines. Installation of native shoreline plants and following Florida Friendly Landscaping principles within living shorelines can help "flip the script" on the impacts on water quality from traditional practices of lawn and landscape professionals, changing their practices while providing new avenues of income for these industries. Florida Sea Grant is working with many partners throughout the state to train various professionals, including those from the landscape industry, on incorporating living shoreline practice into their businesses. (See the sidebar titled "Living Shoreline Training and Education" in chapter 6.)

## Freedom Lawns

When the pesticide and fertilizer applications are stopped and the irrigation is reduced, some non-grass plants that tolerate mowing will also become part of the lawn. Without fertilizers, the total growth rate will slow down, so the lawn will not need to be mowed as often. The best management practice is to mow at the highest setting on the mower—usually 4 to 5 inches (10 to 12 cm)—and to leave most of the grass clippings in place so they can become part of the soil. This higher mowing height means that the grass and other plants in the lawn have more leaf area where photosynthesis can take place, which gives the plants more energy to grow longer roots. The greater height of the lawn plants shades the soil to keep it cooler, and the plants' longer roots also absorb more of the rainfall, so there will be less runoff from the lawn. In North and Central Florida, allowing a lawn to go dormant in the winter reduces the need for mowing.

Many people call these more diverse mowed areas "Freedom Lawns" because they are free from pesticides, free from synthetic fertilizers, free from over irrigation, and free from overseeding in the winter. Since they are mowed less often, they are kinder to our only planet.

**FIGURE 9.1** A freedom lawn is a mixture of plants that tolerate mowing. It is called a freedom lawn because it is free from pesticides, synthetic fertilizers, and over-irrigation. The elimination of landscape-wide applications and reduced mowing is more sustainable and will save money. *Photo by Ginny Stibolt.*

## Stabilizing the Soil

Soil can be eroded by water or by wind, but either way, this soil frequently ends up in the water or wetlands. Eroded soil increases the turbidity of water in marine and estuarine environments, and it also harms wetlands and marshes of all types in several ways. (See chapter 2.) So, stabilizing the soil helps to create a more shoreline-friendly landscape. While lawns, especially freedom lawns, can stabilize the underlying soil to some degree, tightly planted pollinator gardens, established meadows, and thickets or groves of native plants, including trees and shrubs, often do a better job of stabilizing the soil because of their more extensive root systems. In areas where there are other landscape features such as vegetable gardens, pathways, patios, fire pits, or playground equipment, make sure the soil is covered with a combination of pavers, ground cover plants, and mulch that will not float into the waterway during a hard rain. (See more on buffers, mulches, and weed reduction in chapter 8.)

## SOIL ECOSYSTEMS

An amazingly complex ecosystem that most people do not think about is the soil ecosystem, which influences the plants that grow there and the organisms that depend on those plants. One gram of healthy soil (about one-fifth of a teaspoon) can contain 100 million bacteria, 1 million actinomycetes (a special type of bacteria that provide the signature "good-soil" smell), and 100,000 fungi—if strung together, end to end, these microorganisms could measure about 16 feet (4.9 m) in length. This same gram of soil could also contain hundreds of nematodes (which are mostly beneficial), and maybe a few insect eggs or larvae and earthworm cocoons. The exact proportions of these organisms will vary depending on soil conditions such as acidity, moisture, aeration, amount of humus, and the local plant community.[a]

### What Happens Underground

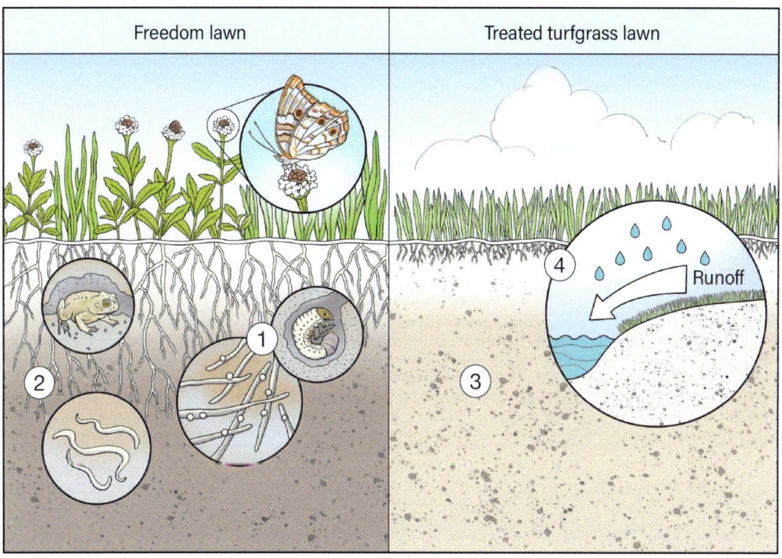


**FIGURE 9.2** The soil under a freedom lawn supports *(1)* fungi, bacteria, nematodes, insects, and various other soil organisms. Healthy soil helps the plants develop *(2)* deeper roots, which absorb more nutrients and soak up more rainwater. Under a turfgrass lawn, which has been treated with fungicides, insecticides and synthetic fertilizers, *(3)* the soil lacks soil organisms that would support the plants. Frequent mowing and applications of synthetic fertilizers cause lawn turf to develop short roots. Soil compaction and short roots *(4)* encourage water to run off, carrying pollutants along with it. *Illustration by Marjorie Shropshire.*

### REBUILDING SOIL ECOSYSTEMS

If soil is sterile subsoil, which is often used to create and fill in new building sites, there may be almost no soil microbes or other soil organisms that normally make up the soil ecosystem. Plants, even native plants, have a much harder time surviving there. Given enough time without synthetic fertilizers and pesticide applications, the soil ecosystem will eventually rebuild itself, but there are ways to aid the process.

The most recommended method to rebuild the soil ecosystem is with an application of a thin layer of finished compost on the top of the soil, including on mowed areas. Caution must be taken for shoreline properties that there is no chance for this nutrient-rich compost to blow off or wash into the water or wetland. Apply it on days with no wind and zero chance of rain. Gently irrigate the area so that the compost is rinsed from the plants and contacts the soil.[b] This should be done at the beginning of the growing season in early spring and, if needed, the process can be repeated several times over a year or two.

It is also a good idea to enrich the soil for newly planted trees and shrubs outside their planting holes to encourage their roots to grow outward so they can become more wind tolerant and more drought tolerant. (See below, page 265.)

[a] Food and Agriculture Organization of the United Nations, "What Is Soil Biodiversity?," accessed January 31, 2025, https://www.fao.org/agriculture/crops/thematic-sitemap/theme/spi/sc-pi-home/managing-ecosystems/soil-biodiversity/soil-what/en/.

[b] Ginny Stibolt, *The Art of Maintaining a Florida Native Landscape* (University Press of Florida, 2015).

## Reducing the Lawn

Another related action to take is to replace at least some of the lawn areas with rain gardens, pollinator gardens, or thickets of understory trees and shrubs planted together or around single trees in the landscape or as hedgerows for screening. In addition, as described in chapter 8, homeowners or property managers can build or enlarge both buffered shorelines and upland buffers. Wooded areas absorb a much higher percentage of the rainfall than lawns. Most lawns, because of their minimal grass root structure and routine soil compaction, can absorb only about 2 inches (5 cm) of rainfall per hour, while a wooded area can handle 14 inches (36 cm) or more per hour.[3] Ideally, most of the water should percolate into the soil and not move across the land surface, which can cause sheet erosion. Reaching this ideal will probably take a few years before the various new landscape plants mature and are able to absorb more stormwater.

## Control Pets

Unrestrained dogs and cats threaten wetland and shoreline habitats and wildlife by destroying nests or preying upon birds and other wildlife. The best way to protect wetland and shoreline wildlife is to keep pets out of the wetland or off the shoreline entirely. Cats can be highly efficient predators and disrupt the natural wildlife cycles and are especially harmful to songbirds.[4]

Pet waste can wash into wetlands and bodies of water, which causes bacterial contamination and completely unnecessary nutrient overload. It is best to keep cats indoors and to keep dogs on a leash or within a fenced-in area where their waste can be removed on a regular basis, especially before rainstorms which could carry the waste with all its nutrients into the wetland or waterway.

## Maintain Septic Systems

Be sure septic systems are operating properly. Faulty or leaking septic systems pollute nearby waterways or wetlands with sewage effluents, which can cause major problems with too many nutrients and possibly toxic bacterial populations such as *E. coli.* This is particularly significant when a property is close to a wetland or a shoreline, because there is little chance for the effluent to be absorbed before it reaches the aquatic ecosystem. If the plants growing over the drain-field are particularly green or if there is a sewage odor during rainy periods, then the system is not working properly. To avoid this problem, have the septic system regularly pumped and inspected—usually every three to five years. If it is an old system (more than twenty years old), or if tree roots have invaded it, it may need to be replaced with a modern system with a new, raised drain-field to pass inspection. Better yet, if there is a local sewer service, hook into that.

## Every Property Is Part of a Watershed

Whether stormwater from a property flows into a local storm drainage system or directly into the adjacent wetland or body of water, it is likely to be detrimental to the water quality in nearby waterways. The US Environmental Protection Agency (EPA) estimates that 85 percent of water pollution comes from nonpoint sources. This includes homeowners' yards, municipal lands, business parks, and community properties.

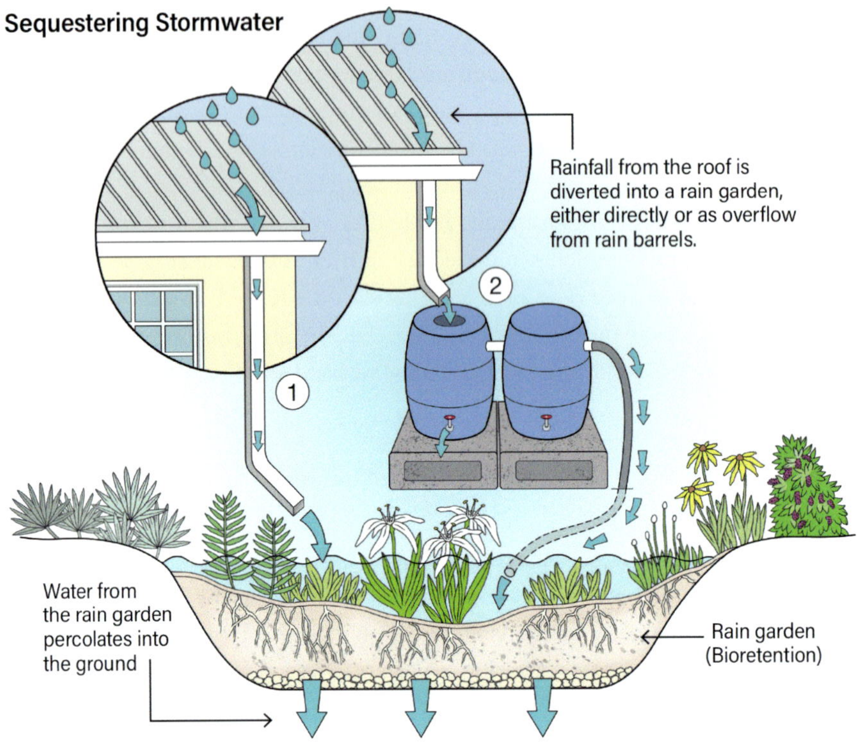


**FIGURE 9.3** Stormwater from the roof with a gutter and downspout can be directed *(1)* into a rain garden where it percolates into the ground or *(2)* into a rain barrel system, which collects rainwater for later use. But when full, the overflow from the second rain barrel is diverted into a rain garden. *Illustration by Marjorie Shropshire.*

There are two main methods of retaining or sequestering rainfall: (1) rain barrels or cisterns, and (2) rain gardens or bioretention areas. The rain barrels or cisterns are normally set up to collect rainwater that falls on roofs of buildings with gutters and downspouts. Rain gardens, bioswales, or other bioretention areas are designed to capture stormwater from impervious surfaces such as driveways, parking lots, and roads, from roofs with gutters and downspouts, overflow from filled rain barrels, or from landscaped or wooded areas. These two actions will make any property less polluting to nearby waterways.

## ▸ GAUGING THE RAINFALL ◂

Knowing the total volume of water for each inch of rain that falls on an impervious surface will help to determine how much capacity the water-sequestration features should have.

The factor for calculating the volume of water falling on an impermeable surface measured in square feet to gallons is 0.62333. For estimates of the capacity of rain barrels or cisterns, 0.6 is close enough. (These calculations do not apply to metric measurements.)

Calculate the roof's surface area (multiply length times width) and then multiply by 0.6 and by the inches of rainfall. (The pitch of the roof or other surface is not a factor in this estimate—measure it as if it were horizontal.)

$$(L \times W \text{ [in feet]}) \times 0.6 \times \text{inches of rain} = \text{gallons of water}$$

For example, 1 inch of rain on 500 square feet of roof will generate 300 gallons of water. This would almost fill six 55-gallon rain barrels.

The average yearly rainfall in Florida is between 39 and 64 inches. If the regional rainfall is unknown for an area, calculate the total by using 50 inches. On average, that 500 square-foot roof will collect approximately 300 × 50 or 15,000 gallons of water each year. That is a lot of water and if it is captured and used for good purposes:

- you do not have to pay for it,
- it will reduce the stress on the local water supply, and most important,
- all that water does not rush directly into the local wetland or body of water.

In Florida, rain gardens should have the capacity to hold at least 2 or 3 inches of rainfall from the collecting surfaces. To size rain gardens, skip the conversion to gallons and just calculate the surface area of the impervious surfaces that feed into that rain garden.

For example, if a 1,000 square-foot roof or impervious driveway feeds into a 5-inch-deep rain garden, a 200-square-foot rain garden will handle 1 inch of rain from that surface. In other words, divide the depth in inches into the total surface area where the rain is collected to calculate the needed area of the rain garden(s). Or do the reverse, if there is a size restriction for a rain garden in that area of the landscape, divide the desired area of the rain garden into the area of the impervious surface to calculate the depth needed to hold 1 inch of water.

$$\frac{\text{(Surface area of collecting surface)}}{\text{(surface area of rain garden)}} = \text{required depth (in inches) of rain garden to absorb 1 inch of rain}$$

For building an endpoint rain garden with extra capacity, say for 4 inches of rain, quadruple the depth in this calculation.

**FIGURE 9.4** A screen-covered intake basket on the top of a rain barrel keeps debris from entering the water, which keeps the barrel cleaner and excludes mosquitoes and other insects. *Photo by Ginny Stibolt.*

## Collecting Rainwater

Gutters catch the rainwater and, all too often in Florida, all that water is piped directly out to the curb, adding to the total runoff, which often carries pollutants and sediment into the local storm drain system and ultimately released directly into nearby waterways. (See chapter 2 for more information on nonpoint source pollution.)

Homeowners, schools, municipalities, churches, and businesses can capture some of the rain that falls on their roofs and then use the collected water in various ways when treated or purified water is not needed. It can be used for washing floors, washing vehicles, irrigation, washing yard tools; prerinsing muddy gardening clothes, muddy hands and feet, muddy pets; and prerinsing freshly harvested root crops such as carrots.

Collected rainfall is particularly important for the irrigation of landscapes, container gardens, and compost piles. Tap water is treated with purifiers to kill bacteria, algae, and mold so we do not get sick, but those purifiers also kill soil microbes found in a healthy soil ecosystem. When collected rainwater is used for irrigation or to wet compost piles, this untreated water does not kill the microbes in soil or compost. Beneficial soil microbes help plants of all types, from those in a vegetable garden to houseplants, stay healthy.

Cisterns are usually installed by professionals and are often configured for green building purposes like flushing toilets. While rain barrels can also be installed by professionals, homeowners or volunteers often build their own rain barrel systems.

## Rain Barrels

There are several sources of rain barrels. Ready-made rain barrels are available for purchase either singly or with hardware to hook several of them together in a series for greater capacity. Many of Florida's county extension offices hold rain barrel workshops, where participants learn how to make a rain barrel and go home with one that they made. This gets people started. Used barrels are often available at bottling, food processing plants, or restaurants, and then they can be converted to rain barrels with detailed instructions from various online resources on how to create and install a rain barrel system that suits the user's specific needs. There is no one right answer, but any method of collecting rainwater is better for the environment than letting it flow unimpeded into waterways or wetlands.

Rain barrels can be set up to work individually or in a series, and rain barrel systems can be either closed or open, as described below. Rain barrels should be mounted in an elevated position so that a bucket or watering can will fit under the hose bibb (spigot), which is inserted near the bottom of the barrel. Mount the rain barrels higher, if there is a need for more gravity pressure to supply adequate water flow for hand watering with a hose or to feed a soaker hose or drip system. The platforms for the rain barrels must be sturdy, because one 50-gallon (190 L) container will weigh more than 400 pounds (180 kg) when full of water.

## An Open Rain Barrel System

There are various configurations for open rain barrel systems, but they all include siphoning off the water from a downspout, and once the barrel(s) are full, the excess flows out into the landscape. At least one barrel in the system needs a method for accepting rainwater from a downspout. This requires a downspout modification designed to get its water into the barrel.

Some systems include plastic or metal hose bibbs (spigots) near the bottom of each barrel so that the barrels fill up one at a time and each barrel drains separately, while other systems attach the barrels via a series of bottom pipe fittings with a single or common spigot so that all the barrels fill and drain as one unit. Open systems include overflows so the excess water can either go into the top of the next barrel in a series or it can empty into water absorbing landscape features such as a rain garden, a seasonal pond, or a natural low space in a wooded area.

## A Closed Rain Barrel System

Closed systems usually consist of a single barrel fitted with a diverter that pushes overflow back into the downspout when the barrel is full. The system is "closed" because there is not a separate overflow outlet—the input and output are through the same pipe or hose. A closed system makes the most sense when the downspout water goes through a pipe away from the building to a drainage system or a water retention feature such as a pond or a rain garden.

The key feature of a closed system is the diverter. When it rains, the diverter directs the rainwater into the barrel through a pipe (or hose) that is mounted on its side near the bottom. When the barrel fills up, the water backs up through the pipe since there are no air holes or vents in the barrel. As the water collects in the catch basin of the diverter, it reaches the top of an overflow pipe and drains into the downspout. There are commercial diverters available for a closed system, but directions to create one are also available online.[5]

## Rain Barrels for Water Sequestration

Once homeowners or property managers have rain barrels, they may find many uses for this "free" water during Florida's seven-month dry season, especially if they grow fruit and vegetables. But to optimize the rain barrels for sequestration and to absorb the first rush of rainfall during a heavy storm during Florida's five-month wet season, it is a good idea to partially empty the barrels between storms, even if the landscape or vegetable gardens do not really need irrigation. To plan for this situation, direct the overflow into an extra-capacity rain garden or seasonal pond. This will allow the water to be absorbed by the rain garden plants or to soak into the soil between those wet-season storm events, and the rain barrels will be emptied enough to contain the first rush of water in the next storm.

## Rain Gardens

Rain gardens are constructed to absorb and filter stormwater in natural or constructed low spots. They are planted with woody and herbaceous plants that absorb much of the water, and they need to be designed so that most of the time, the remainder percolates into the soil within three days, preventing mosquitoes from breeding there. In Florida, where there is a seven-month dry season and a five-month wet season, during which we receive 50 to 70 percent of our annual rainfall, suitable rain garden plants need to tolerate both standing water and drought.

Rain gardens may be designed to work in combination with rain barrels, ponds, or French drains to slow down and manage water flow. A rain garden may be large or small depending upon its location in the landscape, the size of the drainage area, and whether it serves as an endpoint for the stormwater or as an intermediate catch basin. In addition, a rain garden planted with a good variety of native plants also serves as an easy-to-care-for oasis for birds, toads, butterflies, pollinators, and other wildlife.

Roadside swales are a type of rain garden included in many Florida neighborhoods to accept stormwater runoff from roads on one side and from suburban yards on the other side. All too often, the swales are planted with turfgrass, which is difficult to mow as it is an often-soggy environment. These swales are usually on municipal rights-of-way, and there may be regulations as to what can and cannot be planted and how they are to be maintained. The municipality may allow the planting of rushes and sedges (which still allow good visibility for safety) in the bottom of the swale, and those moisture-loving plants will absorb much more water than frequently mowed turfgrass and will tolerate an annual trimming during the dry season to keep trees from taking over. This is a specialized type of rain garden and demonstrates flexibility in considering ways to absorb a greater volume of stormwater.

### Directing Stormwater to the Rain Garden

Rain gardens' stated purpose is to absorb the stormwater runoff from some area or feature in the landscape such as a parking lot, driveway, or roof. Create a path to guide the stormwater into an initial rain garden basin or to the top of the rain garden. Rain gardens can also collect runoff from generally permeable sections of the landscape such as sloping lawns or garden beds. This is usually the case for bioswales used in conjunction with upland buffers. (See chapter 8.)

**FIGURE 9.5** Roadside swales are designed to retain stormwater so that most of it soaks into the ground, but during heavy rainstorms, some may overflow into storm drains or canals and then to larger bodies of water, carrying pollutants with it. *The top two photos are by Vincent Encomio; the lower photo is by Ginny Stibolt.*

If a rain garden is located to absorb the stormwater from a road, parking lot, or driveway, find the tilt and low spot of that surface. Observe the surface in a rainstorm so you will know where the water collects; then arrange to have the rain garden accept the stormwater there. It is generally easiest to build the rain garden right next to the impervious surface, but other configurations may work better for certain landscape features and uses. For instance, a French drain could be installed along the driveway to absorb the stormwater and then empty into a rain garden at a lower spot in the landscape or into a rain garden that is not in the middle of a high foot-traffic area.

An intermediate rain garden is designed to absorb a portion of the stormwater, with the overflow directed into another water retention area such as a wooded area, a seasonal pond, or a drywell. An end-point rain garden should be designed with additional capacity, so that it is unlikely to overflow in a storm.

### DOWNSPOUTS

When gathering runoff from one or more gutter downspouts, sometimes pipes are needed to transport the water underground for several feet or more, depending upon the landscape configuration, and to keep it away from buildings and foot-traffic areas before it empties into the top of a rain garden or series of rain gardens. Install the pipe so there is at least a slight slope away from the building toward the rain garden. Flexible 6-inch (15 cm) plastic drainage pipe is a good choice because it is inexpensive, easy to work with, and requires no special tools. This size is large enough to carry the rainwater from one or two downspouts, even during a heavy Florida rainstorm. When the pipe runs under a path, it is a good idea to center stepping-stones directly over the pipe and support them with firm soil on either side of the pipe, so as people walk on it, their weight is evenly distributed and will not collapse the pipe.

Plan for the rush of water coming from the pipe during a downpour—which could erode the soil in that part of the rain garden—by installing some paving stones or splash rocks to prevent such erosion directly under the pipe outflow. As the rain garden fills up, there may be some backup into the pipe, especially if there's only a slight slope, so it is usually a good idea to drill some holes in the bottom of the pipe as you lay it into its trench or to use perforated pipe to prevent water from puddling. This reduces unseen mosquito breeding areas.

If there's a rain barrel system on the downspout, the last overflow tube can be set up to empty into an underground pipe that leads to water absorbing features. This can be arranged so that the smaller overflow hose hangs loosely into the underground pipe or by a more secure connection, which can be constructed to prevent leaves and animals from entering the pipe.

### FRENCH DRAINS

Modern French drains are usually fashioned using permeable, cloth-covered pipes laid in a ditch next to impervious surfaces or as blind drains that collect water at the overflow end of a rain garden or along a stretch of landscape and deliver it to a water retention site. Set up the configuration to suit the site; just be sure to build in a slight downward angle away from buildings toward a water retention area.

## Sizing a Rain Garden

Determining the required capacity of a rain garden or series of rain gardens depends on how much stormwater will be directed to it and on how quickly the soil drains. Ideally, the rain garden system should have enough capacity to absorb most of the runoff generated in the watershed area. A good capacity for most locations is to handle 2 to 3 inches (5 to 8 cm) of rain. Adjust the rain garden area to get as close as you can to this goal while staying within the space constraints of the site.

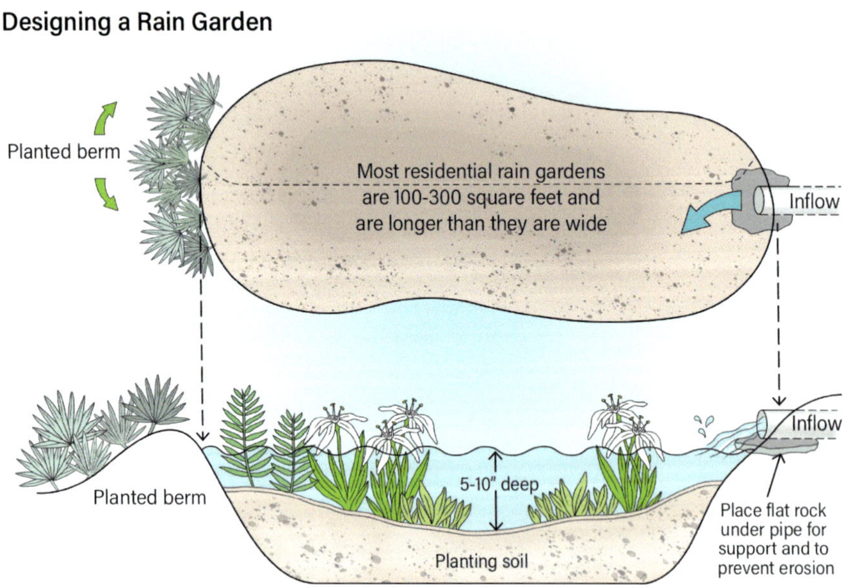


**FIGURE 9.6** Designing a rain garden. Most residential rain gardens are at least 100 square feet (9.3 square meters) and have a flat bottom. A planted berm helps to hold the soil and direct overflow, especially on flat sites. When choosing plants, make sure to account for Florida's dry season—the rain garden will be completely dry at times. The rain garden plants must be able to withstand standing water and drought—a rare combination. *Illustration by Marjorie Shropshire.*

Florida's heavy rains could quickly fill up the garden, so planning for overflow is a must. (See sidebar "Gauging the Rainfall.") The size and the depth of rain gardens will also depend on where they are located on the property and the normal activities that occur in that area.

## Building Rain Gardens

After estimating the surface area and the depth of the rain garden, build rain garden(s) by digging a swale into a level site, by building berms around an area, or by combining these two methods. Normally, the soil dug from the center is used to build the berm. Most residential rain gardens are between 5 and 10 inches (13 to 25 cm) deep, depending upon the slope and configuration of the site. (Large rain gardens in commercial landscapes built to accept runoff from parking lots or roads may be quite deep.) The bottom of the garden should be level, but a berm of soil on the lower edge will catch or contain the water. On a slightly sloping site, the top of that berm should be approximately the same level as the top of the garden. Steeply sloped sites are not suitable for rain gardens, but on moderately sloped sites, a retaining wall on the lower side would provide more capacity and more stability. On a flat site, a berm will be required around the entire garden. More on this below, but for the purpose of sizing the garden or a series of gardens, you will need to estimate the average depth. If the rain garden is in a spot with high foot traffic, a shallower rain garden will create less of a hazard and may be more appropriate. By contrast, a rain garden planted with trees and shrubs will not only absorb more water, but it will also become a feature in the landscape that people can easily see and avoid.

The shape of a rain garden depends on the landscape and how it fits into the space. Generally, it is best to have the broader dimension perpendicular to the slope of the site to capture more of the rainwater from the general landscape as it moves down the slope. If the rain garden (or any other non-lawn feature in the landscape) is installed next to a mowed area, create wide rounded corners and a flat interface space so that the mower can make one clean sweep around the edges and so that no additional trimming will be needed.

## Rain Garden Plants

The plants in a rain garden play an important role by absorbing much of the rainfall directed to it. In general, larger plants with the most leaf surface area will absorb more water than smaller plants with fewer leaves due to their greater rate of transpiration. (For more details on transpiration, see the sidebar "How Plants Work.")

Because it is best to purchase young specimens for planting in the landscape, it will take a few years for the trees and shrubs to gain enough size to absorb large volumes of water.

The seven-month dry season in Florida limits the choices for appropriate rain garden plants since they must tolerate both standing water and drought. While there may be nonnative plants that could possibly be used in a rain garden, the recommended plants on this list are Florida natives. Natives are adapted to the local climate, and they play a crucial role in supporting birds and pollinators. (For more information on native plants, see "Florida's Native, Nonnative, and Invasive Plants.") If the property experiences salt spray, occasional or routine inundation of brackish or saltwater, select plants that will tolerate those conditions as well. Whether plants are recommended for rain gardens is listed in appendix A.

## Planning for Overflow

If a rain garden is to be an endpoint for the stormwater, build some extra capacity—say for 4 to 6 inches (10 to 15 cm) of rainfall and plan for overflows away from buildings. Build a sloped drainage area from the far end of the rain garden toward features that can accept rainwater, such as a drywell, a seasonal pond, more rain gardens, or a low wooded area.

If the landscape is small or the rain garden is in a high foot-traffic area, a drywell at the far end of the garden will add extra capacity for storm water that is not visible and can even be under a path or a lawn. The size of the drywell should correspond with the degree of porosity of the soil: sites with poor drainage will require larger drywells. Line the hole with a couple of layers of geotextile or weed barrier cloth, then fill it with clean gravel or a combination of gravel with cloth-covered perforated pipes or upside-down pots or buckets to create extra hollow spaces where water can collect until it percolates into the soil. Also, make sure the structures in the drywell are sturdy enough so that they will not collapse under the foot or vehicular traffic.

Adding a shallow "seasonal" pond, to which standing water recedes during the drier winter months, provides extra capacity and an overflow area for rain barrels and rain gardens. In many ways this is much like a specialized rain garden. In general, choose a low spot in the landscape, but for shoreline properties, it should be on the upland side of the buffer zone. The vegetation in this pond will most likely cover the whole area. It will be a damp meadow for part of the year and a shallow pond during the wet season. The plants that do well in a rain garden such as native rushes, irises, and sedges will also thrive here. In some locations, a permit may be required for installing a seasonal pond near a shoreline.

**FIGURE 9.7** Steps for building a large drywell that will not be evident in the landscape when it is finished and that will withstand foot traffic. First dig a large hole. (This one is about 4 feet by 6 feet, and 4 feet deep [1.2 × 1.8 m, and 1.2 m deep].) *(A)* Cover 12 foot long (3.7 m) perforated pipes with a woven sock and attach the ends of the pipes together to form a donut. *(B)* Cover the bottom of the hole with geotextile cloth and stack the four pipe-donuts in the hole. The pipes provide more volume for water to be held while it is waiting to percolate into the soil. *(C)* Wrap the outsides of the four pipes with geotextile to keep out the soil. *(D)* Fill the space in the center of the donuts with clean, coarse gravel. (Do not use crusher run gravel with sizes that range from that stated gravel size to grit, because it will not provide enough cavities for the water.) *(E)* Cover the top of the filled donuts with geotextile fabric. More gravel was added to hold the cloth in place to discourage soil-dwelling animals from digging under it. *(F)* Add enough soil on top so that sod that was removed from this space sits at the same level as the surrounding area. *Photos by Ginny Stibolt.*

## Plant More Trees

Preserving mature trees and planting more trees improves both residential and community landscapes in many ways:

- Trees cool the air through transpiration. (See sidebar "How Plants Work.")
- Trees sequester carbon, by absorbing carbon dioxide ($CO_2$), a long-lasting greenhouse gas, through the process of photosynthesis where they remove carbon dioxide from the air to make sugars, which are then stored in the various plant tissues.
- Trees produce oxygen through photosynthesis.
- Trees remove carbon monoxide (CO) and other air pollutants from the air.
- Trees reduce soil erosion with their extensive root systems.
- Trees remove pollutants and sediments from stormwater runoff and then slowly release the water back into waterways and underground aquifers.
- Trees, especially a variety of native trees, provide important habitat for birds and other wildlife.
- Trees increase property values.[6]

### Best Practices for Choosing, Planting, and Caring for Trees and Shrubs

If using container-grown trees and shrubs, it is a good idea to buy small, young plants that are only two or three years old. This not only saves money, but also these younger plants are more likely to survive and adjust to their new locations in the landscape than larger, older specimens. Look for untopped trees with single stems because they will require less corrective pruning and will grow into their natural forms over the years. For shrubs, it is also a good idea to buy younger specimens, but the single stem advice does not necessarily apply, because many shrubs naturally have multiple stems or canes.

When arranging woody plants near a shoreline, build groves with one, three, or five taller trees in the center surrounded by a variety of understory trees and shrubs. (Odd numbers of plants work best for naturalistic arrangements.) Then outside of the shrub layer plant tall bunching grasses or other tall herbaceous plants and surround them with low-growing ground covers. Some of the advantages of this type of arrangement include the following:

- Better wind tolerance because this grouping of plants has a rounded vertical horizon
- Better drought tolerance because the roots of all the plants spread farther out and condition the soil to support mycorrhizal fungi colonization
- Better environment for the plants to grow because of less foot traffic around the roots
- Better habitat for birds and pollinators
- Better absorption of rain and stormwater
- Lower maintenance (eventually)

Once the plants in a grove begin to mature together and create their own environment with their own falling leaves acting as mulch, the maintenance required to keep it looking good will decrease over the years.

After purchasing the initial round of plants, it is a good idea to place all the plants in their containers in the landscape and then to look at the arrangement of the plants from various vantage points including from the house and from other gathering places in the landscape to try to better visualize how they will work together as they grow. Be sure to imagine them as mature specimens and to leave room for the eventual sizes of the woody plants.

Except for broken branches, do not prune any of the woody plants at the time of planting. The trees and shrubs will need lots of energy to grow new roots and to adjust to their new location, and because plants gain energy from photosynthesis in their leaves, there should be no corrective pruning for a year. The next fall, pruning can begin. For trees choose one main leader and remove most suckers (shoots growing straight up from the base of the tree). In this process, remove less than 20 percent of the tree per year. It may take two or three years to correct the initial problems with the plant.

## Rinse the Roots

When planting container-grown trees or shrubs, it is quite likely that they will have been held in containers for several years and that their roots will have grown into a coiling pattern, and if they have been transferred from small containers into larger ones, their coiling pattern might not be evident when the plant is removed from the pot. Sometimes, multiple shrubs might be planted together in one pot to make it look bushier, so when rinsing the roots, separate them so they can be planted individually and farther apart. Root rinsing is not recommended for field-grown or transplanted trees and shrubs; it is recommended only for container-grown specimens.

## Container Grown Plants: How to Plant

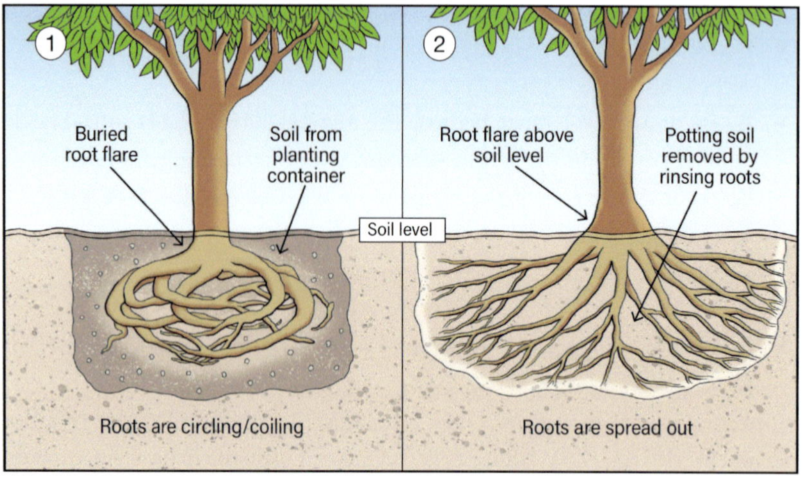


**FIGURE 9.8** *(1)* Planting trees with coiled roots, by including the soil from the planting container and burying the root flare, can lead to roots wrapping around the trunk and eventually choking itself. *(2)* Instead, rinse all the potting soil from the roots, dig a wider planting hole, spread the roots out laterally, and plant in the ground. Make sure the root flare is above the soil level. Do not add any nutrient-rich amendments to the planting hole—just cover the roots with local soil. *Illustration by Marjorie Shropshire.*

## Container Grown Plants: Divide Multiple Plants

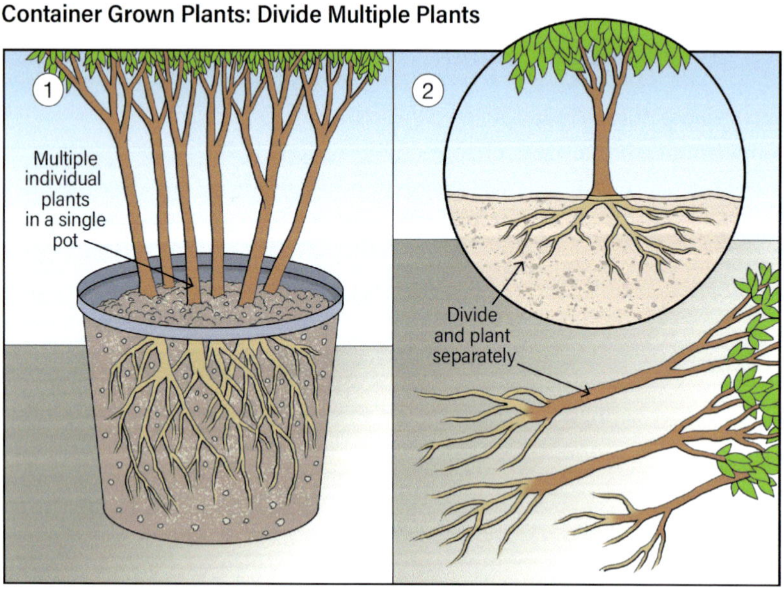


**FIGURE 9.9** *(1)* Some growers plant multiple individuals, especially shrubs, in a single container. *(2)* For more successful growth, separate them and rinse their roots before planting. When the roots are rinsed, the separate plants are easily seen. The trees or shrubs could still be planted as a group, if desired, but with more room for each plant. *Illustration by Marjorie Shropshire.*

If the coiling is not corrected when planted, then years later, those roots could prevent the expansion of the trunk, which would "strangle" the tree or prevent the spreading of surface roots. Either situation could kill the tree or make it less wind tolerant in the future. The best way to reveal the coiling roots is to rinse away all the soil from the root ball so they can be corrected by either uncoiling the smaller roots that are still flexible or by breaking or pruning the roots that are too woody to unbend. Rinsing the roots will also reveal the root flare, where the first main roots flare out from the base of the trunk. The root flare should be at least 2 or 3 inches (5 to 8 cm) above the ground level when the tree is planted. (*Note*: If the tree was grown from a cutting and not a seed, there will not be a root flare per se, but treat those top roots in the same way as a root flare on a seed-grown tree.)

Dig a wide, shallow hole with a firm and higher center where the tree or shrub will be placed so the root flare does not sink below the ground level. No matter how lousy the soil is, do not add any fertilizer, compost, or other enriching amendments to the planting hole. Spread the roots out so they radiate out evenly from the trunk. Some of the roots may need to be staked in place if they are still curved from growing in the container. Cover the roots with 2 or 3 inches (5 to 8 cm) of the local soil, build a berm of soil around the outside of the planting hole, and then flood the hole with water. Flood the hole with water every day for a week or two until the wilting stops, and then irrigate the tree a few days a week for a few more weeks. In the next year, provide some extra irrigation during dry periods. If the tree needs support to stay vertical, stake it for a few weeks until new roots develop—remove the stake(s) as soon as possible so the trunk can build its own strength. After the first few weeks, lay down 2 or 3 inches (5 to 8 cm) of organic mulch, such as wood chips or pine needles, over the planting hole, but at least 6 inches (15 cm) away from the trunk of the tree. No mulch should touch the trunks or stems of the trees and shrubs.

Yes, rinsing the roots and no amendments in the planting hole is harsh treatment, and this may not be the same as older gardening advice, but this method of planting trees has been proven in long-term studies to increase their future viability.[7] The removal of the rich soil in the container and the lack of any enrichments into the planting hole means that the roots are more likely to grow outward, which helps the tree to become more wind tolerant and more drought tolerant—important traits here in Florida.

If the planting site is above the mean higher-high water (MHHW) line and the soil is sandy and lacking in nutrients and organic material, it is a good idea to apply some compost in a circle just outside of the planting hole—do not dig it in, but cover it with mulch. This will enrich the soil and entice the roots to grow outward and this could be repeated one or more times, a month or two

**FIGURE 9.10** Rinsing reveals coiling roots so they can be corrected when planting. This also washes away rich potting soil and fertilizers that were used to keep the plant alive. Rinsing roots like this is a good use of rain barrel water. *Photo by Ginny Stibolt.*

**FIGURE 9.11** Build a berm of soil around the planting hole and flood the hole with water. Continue to water every day for a week or two, and then as needed. The berm will keep water from both irrigation and rain from running off into the surrounding landscape. *Photo by Ginny Stibolt.*

later. The next circle of compost should be laid just outside of the previous one. Enriching the soil below the (MHHW) line is not recommended because those nutrients could be carried into the water during high water events such as storms or king tides.

Also, since there is room between the trees and shrubs (so they will have enough room to grow to their mature sizes) and since young specimens were used, the landscape might look a bit sparse for a few years. So, while waiting for trees and shrubs to grow to maturity, it is a good idea to plant bunching grasses and native wildflowers between them to create a temporary meadow. And then in a few years, as the trees grow and begin to cast significant shade, transplant those meadow plants to sunnier locations and replace them with ferns or other understory plants that can tolerate the shade created by the maturing trees. (See chapter 8.)

**FIGURE 9.12** Native plants surrounding a freshwater retention pond absorb nutrients and moisture and provide habitat for birds and pollinators. The pond was designed with shallow areas around the shoreline to provide planting sites for emergent plants. Notable plants in these fall photos are muhly grasses, bald cypresses, palmettos, and leather ferns. This is one of several water detention and retention sites in a neighborhood adjacent to the St. Lucie River in Martin County. *Photos by Marjorie Shropshire.*

## Freshwater Shorelines

Managing freshwater shorelines is important, and this topic could be another whole book. The regulations are different, the ecosystems are different, and the decision-makers are different. That said, much of what has been covered in this book can also be modified and applied to freshwater habitats.

The shoreline-friendly measures described in this chapter and in chapter 8 would also improve water quality in freshwater habitats, including marshes, swamps, rivers, streams, lakes, or ponds. These actions are most important on properties adjacent to or near these wet habitats, but all properties are in a watershed, and managing stormwater runoff is an important action to improve overall water quality.

To get started, contact the local University of Florida IFAS office and ask about their freshwater programs. "Healthy Ponds" is a new course at the time of this writing, and it helps to educate local officials, homeowners associations, and private citizens about improving the water quality in freshwater ponds and more.[8]

## Shoreline-Friendly Properties Make a Difference

The shoreline-friendly landscaping ideas outlined in this chapter are important measures for all homeowners and property managers, whether they are located next to a body of water or not, because everything that leaves a landscape ends up in a body of water eventually. All property managers can help to improve the water quality of nearby waterways and help recharge the aquifers below. Some projects, like building new upland buffers, as described in chapter 8, may take years to make a significant difference, while others, such as stopping all landscape-wide pesticide and fertilizer applications and sequestering stormwater with rain barrels and rain gardens, will help to improve the water quality right away.

## Notes

1. Mark E. Hostetler and Martin B. Main, "Native Landscaping vs. Exotic Landscaping: What Should We Recommend?" *Journal of Extension* 48, no. 5 (2010): Article 10, https://doi.org/10.34068/joe.48.05.10.
2. Melinda Knuth, Zuan Wei, Zumin Zhang, Hayk Khachatryan, Alan Hodges, and Chengyan Yue, "Preferences for Sustainable Residential Lawns in Florida: The Case of Irrigation

and Fertilization Requirements," *Agronomy* 13, no. 2 (2022): 416, https://www.mdpi.com/2073-4395/13/2/416.

3. Penn State, Department of Ecosystem Science and Management, "For Water Quality: Creating Woods Instead of Lawns," January 7, 2020, https://ecosystems.psu.edu/research/centers/private-forests/news/for-water-quality-creating-woods-instead-of-lawns.
4. Jack Tamisiea, "Cats Kill a Staggering Number of Species across the World," *Scientific American,* December 12, 2023, https://www.scientificamerican.com/article/cats-kill-a-staggering-number-of-species-across-the-world/.
5. Ginny Stibolt, "Rain Barrels Revisited," *Adventures of a Transplanted Gardener* (blog), 2009, http://www.sky-bolt.com/garden/rainbarrels2.htm.
6. Ginny Stibolt, "The Value of Trees," *Green Gardening Matters* (blog), April 22, 2022, https://greengardeningmatters.blogspot.com/2022/04/the-value-of-trees.html.
7. Linda Chalker-Scott. *Horticultural Myths* (blog). Accessed February 6, 2025. https://https://puyallup.wsu.edu/lcs/.
8. Lee Hayes Byron, Armando Ubeda, Carol Ywatt-Evens, and Michael D'Imperio, "Resources for Neighborhoods in Protecting Water Quality," University of Florida, IFAS Blogs, November 9, 2022, https://blogs.ifas.ufl.edu/sarasotaco/2022/11/09/resources-for-neighborhoods-in-protecting-water-quality/.

## References

Chalker-Scott, Linda. *Horticultural Myths* (blog). Accessed February 6, 2025. https://https://puyallup.wsu.edu/lcs/.

Stibolt, Ginny. "Green Resources." *Green Gardening Matters* (blog), accessed February 6, 2025, https://greengardeningmatters.blogspot.com/p/green-resources.html.

Stibolt, Ginny. *Sustainable Gardening for Florida.* University Press of Florida, 2009.

# Epilogue

## *The Ripple Effect of Living Shorelines*

**AS WE CLOSE OUR** journey through the world of living shorelines, it is important to remember that they are just one example of a broader "green infrastructure" or "nature-based solutions" movement. This book has covered many examples of green infrastructure for individual shorelines and yards, but there are many larger-scale types of green infrastructure that can be employed on public lands and municipal properties. By embracing green infrastructure and becoming living shoreline ambassadors, individuals hold the power to create substantial environmental change, making communities more resilient to the growing threats of sea level rise, extreme precipitation and drought, and increasing temperatures.

Homeowners, in particular, wield significant influence. Through their choices and preferences, they can guide their contractors toward sustainable practices. As the client, the homeowner is in a position to request a living shoreline over an armored solution. The decision whether to opt for a living shoreline not only influences the team of professionals working at the site but also serves as a living example for others to follow. A living shoreline can be a great conversation starter for communities and neighbors looking for innovative and cost-effective[1] strategies to address shoreline management.

Research supports the idea that influence extends beyond property lines.[2] The decision to armor a shoreline can cascade into additional armoring in a community. Likewise, the choice to stick with a natural shoreline or to install a living shoreline can inspire neighbors, nudging them toward more environmentally responsible shoreline management decisions. This domino effect can transform entire neighborhoods, turning them into bastions of ecological resilience.[3]

In addition to individual neighbors, living shoreline ambassadors can have great influence in community homeowners' associations and at local governance meetings. By actively participating in public meetings and discussions, you can suggest green infrastructure as an avenue to address flooding or erosion issues, voice support for proposed green infrastructure projects, and advocate for policies that support green infrastructure. Individual engagement is crucial in shaping the environmental agenda of any community, county, or town.

In today's interconnected world, the reach of one person's voice can be amplified by social media and local news platforms. Use these tools to share experiences, educate others about living shorelines, and spread awareness about their benefits. Stories, pictures, and posts can go a long way in inspiring others and garnering support for nature-based initiatives.

**FIGURE 10.1** The living shoreline stakeholder ecosystem. The health and well-being of Florida's people, animals, plants, environment, and economy all benefit from living shorelines. *Illustration by Marjorie Shropshire.*

As this book concludes, remember that the journey toward a sustainable future continues. The steps we take today toward adopting living shorelines and other green infrastructure can set the foundation for a healthier planet tomorrow. The role of an ambassador for the environment is critical. So, speak up, take action, and inspire change. The future is up to all of us, and with each small wave of change, we can create an ocean of difference.

## Notes

1. Camille Sicangco, Renee Collini, Sara Martin, Alexis Monti, and Eric Sparks, *Cost-Benefit Analysis of a Small-Scale Living Shoreline Project* (Mississippi-Alabama Sea Grant Publication MASGP-21–054, 2021), https://placeslr.org/wp-content/uploads/2021/07/Living-Shoreline-Cost-Benefit-AnalysiswithPub.pdf.
2. Rachel K. Gittman, Steven B. Scyphers, Christopher J. Baillie, et al., "Reversing a Tyranny of Cascading Shoreline-Protection Decisions Driving Coastal Habitat Loss," *Conservation Science and Practice* 3, no. 9 (2021): e490. https://doi.org/10.1111/csp2.490; Savanna C. Barry, Laura K. Reynolds, Anna E. Braswell, Rachel K. Gittman, Steven B. Scyphers, and Ashley R. Smyth, "Perceived Effectiveness Drives Shoreline Decision-Making for Florida's Waterfront Property Owners," *Ocean and Coastal Management* 258 (2024): 107353, https://doi.org/10.1016/j.ocecoaman.2024.107353.
3. Mariko A. Polk, Rachel K. Gittman, Carter S. Smith, and Devon O. Eulie, "Coastal Resilience Surges as Living Shorelines Reduce Lateral Erosion of Salt Marshes," *Integrated Environmental Assessment and Management* 18, no. 1 (2020): 82–98, https://doi.org/10.1002/ieam.4447.

# Appendix A

# Suggested Plants

**THIS LIST OF FLORIDA** native plants includes those that are easy to grow and generally available from the Florida native nurseries. There are two parts to this list: salt-tolerant and non-salt-tolerant plants. Each list is arranged by form: trees and shrubs, palms, forbs (herbaceous flowering plants), ferns, graminoids (grasses, rushes, sedges, and other grass-like plants), and vines. If a plant is a primary food for butterfly or moth larvae, it is listed as a "Butterfly host plant." This list is just a starting point; there are other more complete plant references—both in print and online. For purposes of this book, the botanical authority for accepted names and whether plants are native or not is R. P. Wunderlin, B. F. Hansen, A. R. Franck, and F. B. Essig's 2025 *Atlas of Florida Plants,* hosted by the University of South Florida (https://florida.plantatlas.usf.edu/).

The following references will offer further help in choosing the best plants for any Florida landscape:

- Gil Nelson, *Florida's Best Native Landscape Plants: 200 Readily Available Species for Homeowners and Professionals* (University Press of Florida, 2003).
- The Florida Native Plant Society has a tool on its website for finding natives for landscaping. Select by county or zip code the conditions and properties desired, such as salt tolerance or providing wildlife food and cover. This tool will generate a list of suggested plants sorted by plant form. (https://www.fnps.org/plants)

- For South Florida, the website of the Institute for Regional Conservation, "Natives for Your Neighborhood," provides a tool to find appropriate native plants by zip code. (https://www.regionalconservation.org)
- Specifically for living shorelines, the Florida Living Shorelines website (https://floridalivingshorelines.com/) lists appropriate native plants by tidal zone. Navigate to the "Explore by Region" page within the website and select the appropriate Florida region. Please note that the combined "Mid-Marsh" and "High Marsh" zones defined on the website are synonymous with our description of high marsh in chapter 7.
- Building a relationship with local native plant nurseries that can help make the right decisions for specific sites and to ensure that plants are bred from local stock is an important step in the success of living shoreline projects. In Florida, go to Plant Real Florida (https//www.plantrealflorida.org) and the Florida Association of Native Nurseries (https://www.fann.org/) to find local native nurseries and nurseries that have a particular species in stock.

## Salt-Tolerant Plant List

### Trees and Shrubs

**Agaves** (*Agave* spp.) (zones 6–11). There are two species native to Florida. False sisal (*A. decipiens*) is native to the coastal regions in South Florida and is tolerant of salt spray and occasional inundation with brackish water. It often occurs in the upland regions of salt marshes and mangrove forests. False aloe (*A. virginica*) has a wider range (from Florida's Panhandle north to Virginia and south through eastern Mexico), and it is not restricted to coastal areas and is apparently not particularly salt tolerant. Both species grow to 5 feet (1.5 m) or more before they bloom and (like any agave, including century plants) they die after blooming, usually producing pups, which can be replanted to continue the population. Butterfly host plant. Maintenance: No particular care is needed once established. They are extremely drought tolerant.

**Bay cedar** (*Suriana maritima*) (zones 9–11). This evergreen shrub, with showy yellow flowers and fleshy leaves, grows to 6 to 10 feet (1.8 to 3 m) tall and is usually as broad as tall with its many ascending branches. It is native to coastal regions in Central and South Florida and grows well in sandy, nutrient-poor soils. It is drought tolerant, tolerant of direct salt spray and occasional inundation by saltwater.

Butterfly host plant. Maintenance: No particular care is needed once established.

**Beachberry** (*Scaevola plumieri*) (zones 9–11). This small evergreen shrub has thick, succulent leaves and trailing stems. It grows to about 2 to 5 feet (0.6 to 1.5 m) in height and is normally broader than tall. Its white flowers have five petals, but only on one side. The fruits are black. It is native to coastal areas in South and Central Florida. It is drought tolerant and tolerant of significant salt spray, but it is not tolerant of inundation by brackish or saltwater. Maintenance: No particular care is needed once established.

**Beautyberry** (*Callicarpa americana*) (zones 6–11). This deciduous shrub lives up to its name with its unbelievably vivid purple berries that are eaten by a wide variety of birds. It tolerates some salt spray, but not inundation of brackish or saltwater. It grows well in full sun or partial shade, and it is not fussy about soil. The gracefully arching canes can reach out to more than 6 feet (1.8 m) from the main plant. Maintenance: It reseeds to some extent, so some seedlings may need to be transplanted if there are too many.

**Black torch** (*Erithalis fruticosa*) (zones 10–11). This evergreen shrub grows from 4 to 8 feet (1.2 to 2.4 m) tall and nearly as wide. It blooms and fruits year-round and is native to Southeastern Florida and the Keys. It is drought tolerant, tolerant of salt spray, and tolerant of occasional inundation of salt or brackish water. Maintenance: No particular maintenance is required once they are established.

**Blueberries**, sparkleberries, and deerberries (*Vaccinium* spp.) (zones 5–11). The five species in this genus native to Florida are shrubs that can be tall or short. Sparkleberry (*V. arboretum*) and shiny blueberry (*V. myrsinites*) are tolerant of salt spray, but none of them is tolerant of inundation by brackish or saltwater. They prefer acidic soils and full sun to partial shade. All provide good habitat value. Maintenance: Unless they are planted specifically as edible crops, no particular maintenance is required.

**Buttonwood** (*Conocarpus erectus*) (zones 9–11). A small tree that naturally occurs in coastal areas at the edges of fresh, brackish, or saltwater marshes; it has two natural color morphs—green and silver. It grows best in full sun or partial shade. It is useful for building upland buffers in Central and South Florida (see chapter 8), but it can also withstand the rigors of urban settings and makes a durable street or parking lot tree. Butterfly host plant. Maintenance: No particular care is needed once established, but it tolerates trimming if necessary.

**Cocoplum** (*Chrysobalanus icaco*) (zones 10–11). This dense evergreen shrub grows in a variety of conditions from swamps to beaches. It is tolerant of salt spray and occasional inundation by brackish water. It does best in rich soil in a sunny location, and it is recommended for hedgerows. It bears fruit mostly in the winter but continues intermittently throughout the year. The fruit is noticeable and eaten by wildlife and humans alike. Maintenance: Germination of seeds is slow, but once established, no particular maintenance is required.

**Coontie** (*Zamia integrifolia*) (zones 8–11). This is Florida's only native cycad (a primitive nonflowering plant with male and female individuals). It grows slowly and is not a real shrub since it does not have wood, but it serves as a tidy shrub in the landscape. It can grow in full sun or partial shade, and it is drought tolerant. It is also tolerant of salt spray and inundation by brackish water. Butterfly host plant. Maintenance: Keep vines and aggressive weeds away.

**Coral bean** (*Erythrina herbacea*) (zones 8–11). This perennial dies back to the roots in areas that have frost, but it can become a small tree in frost-free regions. It is a drought-tolerant legume with showy red flowers and red poisonous fruit. It grows in dry rich soils, but it adapts to a wide range of habitats as long as they are well drained. It is tolerant of salt spray but not of inundation with brackish or saltwater. The flowers attract pollinators, including hummingbirds; and birds and other wildlife consume the fruit. Maintenance: No particular care is needed.

**Christmasberry** (*Lycium carolinianum*) (zones 8–11). This evergreen shrub has needle-like succulent leaves and can grow to about 6 or 8 feet (1.8 to 2.4 m) tall, and it grows as broad as it is tall. It does best in moist or wet soil and full sun, and it is wind tolerant. It is tolerant of salt spray and regular inundation with brackish or saltwater. It is found along salt marsh and mangrove edges and on dunes in Florida's coastal counties. The flowers are lavender or bluish and attract a wide variety of pollinators; its fruit is bright red like a tomato, and indeed it is in the same family (Solanaceae or nightshade). Reports on edibility of the fruit vary, but many birds feed on the fruit. (*Note:* Another Florida native plant also known as Christmasberry (*Crossopetalum ilicifolium*) occurs in South Florida and is not salt tolerant.) Maintenance: No particular care is needed once established.

**Cypresses** (*Taxodium* spp.) (zones 6–11). Bald cypress (*T. distichum*) and pond cypress (*T. ascendens*) are large deciduous conifers that occur naturally in wet, acidic areas in all of Florida, but the bald cypress can adjust to drier habitats. In damp areas, they tend to develop knees (gnarly, vertical root extensions) that may impede walking and mowing, even at some distance from the tree. Both species are wind tolerant, and the pond cypress is also tolerant of salt spray and occasional inundation by brackish water. Either would work as a rain garden tree. Butterfly host plants. Maintenance: No particular maintenance is necessary.

**Florida swamp privet** (*Forestiera segregata*) (zones 8–11). This large drought-tolerant, evergreen shrub can be used as part of a hedgerow. It grows best in moist rich soil, but it will adapt to other habitats. It is tolerant of salt spray and of occasional inundation by brackish or saltwater. It attracts pollinators and birds. Butterfly host plant. Maintenance: It tolerates trimming well, but if allowed to grow out to its natural size, no particular maintenance is needed.

**Gopher apple** (*Geobalanus oblongifolius*) (zones 8–11). This slow-growing woody subshrub stays close to the ground in poor soil with a maze of underground stems. It could be used as a ground cover and turfgrass substitute in hot sandy places. It is tolerant of salt spray but not of inundation by brackish or saltwater. Maintenance: High mowing annually to keep down weeds while it becomes established.

**Gumbo limbo** (*Bursera simaruba*) (zones 10–11). Plant as a specimen tree to show off its bronze-colored peeling bark. It also makes a good shade tree. It is tolerant of salt spray and occasional inundation by brackish or saltwater. Butterfly host plant. Maintenance: Its normal rounded shape reduces the need for pruning if given enough room. No particular care is needed, once established. If a branch breaks off, it will grow a new tree if stuck in the ground.

**Hollies** (*Ilex* spp.) (zones 4–11). More than a dozen hollies are native to Florida; most are not salt tolerant, but Krug's holly (*I. krugiana*) (zone 11) tolerates salt spray but not inundation; and yaupon (*I. vomitoria*) (zones 7–11) tolerates salt spray and inundation by brackish water. Most hollies are evergreen, but winterberry (*I. verticillata*) is deciduous, leaving its striking red berries on bare twigs in the winter. Most hollies are understory trees, but some like gallberry (*I. glabra*) are shrubs. Hollies are dioecious, so make sure at least one plant is male so the females can produce those famous fruits. Hollies grow in a wide variety of habitats. Most hollies are single trees but some, especially yaupon holly (*I. vomitoria*), send up lots of suckers and will spread. Maintenance: Even though most are evergreen, hollies will drop their leaves.

**Jamaican capertree** (*Quadrella jamaicensis*) (zones 9–11). A tall, broad, evergreen shrub or small tree for South Florida that is both drought tolerant and tolerant of salt spray and occasional inundation by brackish or salt water. It is recommended for hedgerows in areas with neutral to alkaline sandy soils, especially in coastal areas. Its beautiful blooms change from white to pink. Butterfly host plant. Maintenance: Its normal rounded shape reduces the need for pruning if given enough room. No particular care is needed, once established.

**Magnolias** (*Magnolia* spp.). There are six species native to Florida and these evergreen trees prefer acidic soil and can grow in full sun or partial shade. They all bear white showy flowers, and their bright red fruit is borne on a cone-like fruiting structure. Birds eat the fruit. The two species that are the most prevalent and the most available are also the ones with at least some salt tolerance. Southern magnolia (*M. grandiflora*) (zones 7–9) is tolerant of salt spray but not of inundation with brackish or saltwater. Sweetbay (*M. virginiana*) (zones 5–11) prefers moist or wet sites and tolerates periodic inundation, so it is a good choice for rain gardens. It tends to send up suckers that form thickets in damp places. It is tolerant of salt spray and also of inundation by brackish or saltwater. Butterfly host plants. Maintenance: Magnolia leaves are large, so keeping them cleared may be a problem if planted near mowed areas.

**Mangroves** (zones 9–11). Mangroves occur mainly in intertidal areas with brackish or saltwater where they play an important role in controlling coastal erosion in intertidal areas, shorelines, and uplands. In fact, the peat soils under mangroves tend to keep up with sea level rise and sequester an outsized portion of carbon. They provide good structural habitat for birds and aquatic animals. Black mangrove (*Avicennia germinans*) is a subtropical, salt-tolerant shrub in its northern range, but a medium-sized tree in frost-free areas. It is the most cold-hardy of the three true mangroves in Florida. Red mangrove (*Rhizophora mangle*), a salt tolerant shrub or medium-sized tree has stilt-like roots and grows farthest out in the water compared to other mangroves. It will grow in either saltwater or freshwater habitats, but it mostly occurs in salt water. White mangrove (*Laguncularia racemosa*) is a bit less cold tolerant and has not migrated as far north as the black and red mangroves. When growing with other mangrove species, the white mangrove usually occurs on the landward side of the other two. Butterfly host plants. Maintenance: No particular care is needed once established, and in fact, there are trimming regulations. See chapter 1 for more information on mangroves.

**Marlberry** (*Ardisia escallonioides*) (zones 9–11). This evergreen, shade-tolerant small tree is attractive to pollinators and birds. It occurs naturally as an understory plant, but it will work as a screen or in a hedgerow. It is tolerant of salt spray but not of inundation by brackish or saltwater. (Do not confuse it with the invasive coral ardisia [*Ardisia crenata*].) Maintenance: It has brittle wood, so do not plant it in highly exposed areas; some pruning may be needed after high-wind events.

**Marsh elders** (*Iva* spp.) (zones 6–11). Four species of marsh elders are native to Florida—all are low-growing, drought-tolerant shrubs. They are tolerant of salt spray and inundation by brackish or saltwater. They grow naturally on sand dunes and are excellent choices for beach stabilization. They spread by deep runners, so leave plenty of room for them to spread out. Maintenance: No particular maintenance is required—they stay low.

**Necklace pod** (*Sophora tomentosa*) (zones 9–11). An easy-to-grow tree or shrub with a long bloom cycle. It is tolerant of occasional/brief inundation such as can occur in storm surges. It can tolerate significant and ongoing amounts of salty wind and salt spray without injury. Maintenance: No particular maintenance is required.

**Oaks** (*Quercus* spp.) (zones 3–11). Most of the more than twenty oak species native to Florida are not salt tolerant, but white oak (*Q. alba*), Chapman's oak (*Q. chapmanii*), sand live oak (*Q. geminata*), scrub oak (*Q. inopina*), myrtle oak (*Q. myrtifolia*), willow oak (*Q. phellos*), and post oak (*Q. stellata*) are tolerant of salt spray but not of inundation by brackish or saltwater. The live oak (*Q. virginiana*) is tolerant of both salt spray and occasional inundation by brackish or saltwater. The oaks prefer acidic upland soils and range in size from large (80 feet [24m] tall and 130 feet [40m]) wide for the live oak, *Q. virginiana*) to small (less than a foot tall) for the dwarf live oak (*Q. minima*). Some grow quickly, but most do not, so choose carefully. Oaks provide excellent wildlife habitat value. Butterfly host plants. Maintenance: Most oaks reseed profusely. In the fall and winter most oaks lose their leaves which do not decompose quickly, so avoid planting lawn areas around these trees.

**Persimmon** (*Diospyros virginiana*) (zones 6–11). This small to medium-sized deciduous tree is dioecious, with the females bearing good-sized fruits that are eaten by wildlife and humans alike. Be sure to include at least one male plant to produce fruit. It is adaptable to various habitats and has good fall color. It is tolerant of salt spray but not of inundation by brackish or saltwater. It is best planted at edges of wooded areas. Insect pollinated. Butterfly host plant. Maintenance: Fallen fruit attracts wasps and can be messy, so plant where this will not be a problem, or harvest the fruit as it ripens.

**Pines** (*Pinus* spp.) (zones 3–11). Only two of the six pines native to Florida have any salt tolerance. Sand pine (*P. clausa*) is tolerant of salt spray but not of inundation by brackish or saltwater. Slash pine (*P. elliottii*) is tolerant of salt spray and occasional inundation by brackish water, but not saltwater. The pines prefer acidic upland habitats and are often the pioneer woody plants in fields and meadows. Most pines, such as the longleaf pine (*P. palustris*), are adapted to fire, but they and their pine needles are highly flammable. Butterfly host plants. Maintenance: Avoid pines in lawn areas because the needle cleanup can be an ongoing chore. Plant away from buildings in fire-prone areas.

**Pond apple** (*Annona glabra*) (zones 10–11). This is a small to medium tree or shrub with a rounded crown from wide-spreading branches with thick, short, and often buttressed trunks. It is moderately tolerant of salt spray and of brackish water inundation with only occasional saltwater surges. It is native to aquatic habitats of South Florida and some coastal regions of Central Florida, but it adapts to somewhat drier conditions. Butterfly host plant. Maintenance: The fruit drops can be messy, but other than that, no particular maintenance is necessary.

**Red cedar** (*Juniperus virginiana*) (zones 3–10). This medium-sized, evergreen conifer is a pioneer species that grows well on upland disturbed sites, at first growing into a perfect elongated cone shape. As it matures the crown becomes rounded or irregular. It is often used as a screen; it is a good addition to a hedgerow, but give it room to spread. It is tolerant of salt spray and occasional inundation by brackish or salt water. It is wind tolerant. It offers good habitat value with its denseness and its waxy, berrylike cones. Maintenance: No particular maintenance is needed.

**Salt bush,** or groundsel tree (*Baccharis halimifolia*) (zones 3–9). This dioecious, semi-evergreen shrub grows just about anywhere, but it does better in moist habitats. It is tolerant of salt spray and occasional inundation by brackish water. The female individuals have showy white flowers with fluffy seeds in the fall. It is a good choice for a hedgerow. Butterfly host plant. Maintenance: It takes to trimming if necessary. Reseeds freely but it is relatively short-lived, so let some of those seedlings grow to keep a good population. The wood is brittle and may require pruning after high-wind events.

**Saltwort,** or turtleweed (*Batis maritima*) (zones 7–11). This succulent low-growing shrub (2 to 4 feet (0.6 to 1.2 m) tall) has spreading branches, which often root at the tips. Its native range is coastal areas of southern North America, Central America, and northern South America. It covers large areas of tidal salt marshes and is tolerant of salt spray and frequent inundation by salt water. Butterfly host plant. Maintenance: Not drought tolerant, but if planted where it is wet and where they can spread, no particular maintenance is necessary.

**Sea lavender** (*Heliotropium gnaphalodes*) (zones 9–11). This is a mounding shrub that grows up to 6 feet (1.2 m) tall with succulent, silver-gray foliage. It naturally occurs on sand dunes and prefers sandy, well drained soils. It is tolerant of salt spray and only occasional inundation or surges of saltwater. Also see other species of this genus under salt-tolerant forbs. Maintenance: If planted where they can spread, no particular maintenance is necessary.

**Sea oxeyes** (*Borrichia* spp.) (zones 6–11). There are two species native to Florida. Bushy seaside oxeye (*B. frutescens*), which is a shrub with silvery leaves that is about 3 or 4 feet (0.9 to 1.2 m) tall that is native to all of Florida's coastal counties and tree seaside oxeye (*B. arborescens*), which grows is up to 6 feet (1.8 m) tall and is native to only the southernmost counties in Florida. The sea oxeyes are both highly drought- and salt-tolerant and can withstand occasional saltwater inundation. They both can form colonies. They have yellow flower heads that attract butterflies and other pollinators, and they can bloom year-round. Its seeds provide food for birds and other small wildlife. Maintenance: If planted where they can spread, no particular maintenance is necessary.

**Sea grape** (*Coccoloba uvifera*) (zones 10–11). This small tree or large shrub is evergreen and salt tolerant. It grows in poor sandy soil and is also drought tolerant.

It is dioecious; its showy grapelike clusters of berries are borne on the female plants. It usually has irregularly shaped multiple stems and is often seen around sand dune areas used as a hedge and windbreak. Maintenance: Its large leathery leaves will need to be cleaned up on a regular basis from walkways and lawns, so plant where the leaf drop does not matter.

**Simpson's stopper** (*Myrcianthes fragrans*) (zones 9–11). This slow-growing evergreen shrub or small tree is suitable for hedges or hedgerows. It tolerates alkaline soils. It is drought tolerant but will do best if irrigated during dry periods the first few years after planting. Its fragrant flowers and colorful fruit are attractive to wildlife and humans. It is somewhat tolerant of salt spray but not of inundation by brackish or saltwater. Maintenance: No particular maintenance is necessary after establishment.

**Sumacs** (*Rhus* spp.) (zones 4–11). There are four species native to Florida. They are deciduous, dioecious, pioneer species that often form thickets or colonies. They can grow to more than 20 feet (6 m) tall but stay short in poor soils. Found mostly in sandy acidic soils in full sun or partial shade. The showy greenish-white flowers attract many pollinators; later, flowers on the female shrubs will mature into red fruits. It has good fall color, even in Florida. The winged sumac (*R. copallinum*) is the only one that is tolerant of salt spray, however, it does not tolerate inundation by brackish or salt water. (*Note:* These are not poison sumac [*Toxicodendron vernix*].) Butterfly host plant. Maintenance: They spread via rhizomes, so if it produces suckers in unwanted spaces, it is easy to pull them up or transplant them. If planted where it can spread, no particular maintenance is necessary.

**Sweet acacia** (*Vachellia farnesiana*) (zones 8–11). This spiny, evergreen member of the bean family is easy to grow and produces showy yellow flowers and beanlike fruits. It is tolerant of occasional, brief inundation by seawater, such as can occur in storm surges, and it can tolerate significant and ongoing amounts of salty wind and salt spray without injury. Maintenance: No particular maintenance required.

**Sweetpepperbush**, coastal sweetpepperbush (*Clethra alnifolia*) (zones 3–9). This deciduous understory shrub prefers slightly acidic moist sites and will form thickets, particularly in wetter areas. It is tolerant of salt spray and occasional inundation by brackish or saltwater. Its highly fragrant flowers attract a wide variety of pollinators, including hummingbirds. Maintenance: No particular maintenance required.

**Sweetgum** (*Liquidambar styraciflua*) (zones 6–10). This large tree thrives in many habitats, but it may grow faster in moist sites. It is somewhat tolerant of salt spray, but not of inundation by brackish or saltwater. Butterfly host plant. Maintenance: Produces abundant, hard, spiny seed pods (gumballs), and it readily reseeds and sends up numerous root suckers at some distance from the tree.

**Tupelos** (*Nyssa* spp.) (zones 4–10). The four native tupelos are small to medium-sized deciduous trees, often with an interesting, gnarly growth pattern. They have good fall color even in the southern ranges. They prefer moist or wet slightly acidic soils. The flowers attract pollinators and birds, and other wildlife consume the berries. Tupelos are good contrast trees in the landscape. They are all good rain garden plants. Black gum (*N. sylvatica*) is tolerant of salt spray but not of inundation by brackish or saltwater. Maintenance: No particular maintenance is needed. (*Note:* On young trees, the branches are perpendicular to the trunk and may interfere with nearby foot or vehicular traffic, so be sure to allow space for this growth pattern.)

**Varnish leaf** (*Dodonaea viscosa*) (zones 9–10). This evergreen shrub typically grows from 10 to 15 feet (3 to 4.5 m) tall in coastal upland areas in most of peninsular Florida, and it is tolerant of salt spray and occasional inundation by salt or brackish water. Small-fruited varnish leaf (*D. elaeagnoides*) grows to about 6 feet (1.8 m) tall and occurs in similar environments, but only in the Keys. Maintenance: No particular maintenance is needed.

**Wax myrtle, southern bayberry** (*Morella cerifera*) (zones 7–11). This highly adaptable, evergreen shrub or small tree grows well in poor soils because its roots fix nitrogen. It is tolerant of salt spray and inundation by brackish or saltwater. Dioecious. It provides food for winter songbirds with its prolific waxy berries borne on female plants. It is good for screening or hedgerows. Rain garden plant. There are two other *Morella* species: odorless bayberry (*M. inodora*) and northern bayberry (*M. caroliniensis*), but neither is salt tolerant. Butterfly host plant. Maintenance: It produces suckers, so plant where that will not make a difference. For the least amount of maintenance, allow space for its adult size, but it does tolerate trimming to create a tidy hedge.

**White indigoberry** (*Randia aculeata*) (zones 9–11). This thorny evergreen shrub grows from 3 to 10 feet (1 to 3 m) tall and bears white flowers and white fruit. It grows naturally in upland locations in coastal areas in Central and South Florida. It is tolerant of salt spray but not of frequent inundation by brackish or saltwater. Butterfly host plant. Maintenance: No particular maintenance is needed for this drought-tolerant shrub.

**Wild coffee** (*Psychotria nervosa*) (zones 9–11). This evergreen understory shrub tolerates full shade. Its attractive fruits and shiny leaves create interest in the landscape and provide contrast in a hedgerow. It prefers well-drained neutral to alkaline soils. It is somewhat tolerant of salt spray but not of inundation by brackish or saltwater. Attracts pollinators. There is also a dwarf variety. Maintenance: No particular maintenance is needed.

**Wild lime,** or lime pricklyash (*Zanthoxylum fagara*) (zones 9–11). This thorny, small to medium, evergreen, aromatic tree prefers poor, alkaline soil but is adaptable. It can be used as a screen or in a hedgerow. It offers good habitat value and has interesting foliage. It is tolerant of salt spray and occasional inundation by brackish or salt water. Butterfly host plant. Maintenance: It naturally produces multiple trunks, but they tolerate pruning some of them out for looks or airflow.

**Wild sage** (*Lantana involucrata*) (zones 9–11). This shrub bears its white flowers year-round. Tolerant of occasional/brief inundation such as can occur in storm surges and some tolerance to salty wind but not direct salt spray. (*Note:* This is not the invasive *Lantana camara*.) Maintenance: No particular maintenance is needed.

**Yellow necklacepod** (*Sophora tomentosa* var. *truncata*) (zones 9–11). This shrub in the bean family can grow in poor soils since it fixes nitrogen in its roots. It grows to 6 or 8 feet (1.8 to 2.4 m) in height and occurs naturally in coastal areas throughout Central and South Florida. Its showy yellow flowers attract butterflies and hummingbirds throughout the year. It has a high tolerance for salt spray and is tolerant of only occasional inundation of saltwater. (*Note:* There is a non-native species also known as yellow necklace pod [*S. tomentos*a var. *occidentalis*], which is often sold in big box garden centers. The non-native species is covered with fine hair. The two species can cross-pollinate, so care should be taken to avoid the non-native variety.) Maintenance: No particular care is needed, but its bean-shaped seed pods are poisonous so do not plant where that would be a problem.

**Yuccas** (*Yucca* spp.) (zones 4–11). There are three species native to Florida. Both Spanish bayonet (*Y. aloifolia*) and Adam's needle (*Y. filamentosa*) are tolerant of salt spray and occasional inundation by brackish or saltwater, but Spanish dagger (*Y. gloriosa*) is not tolerant of salt spray or inundation. Yuccas offer a spiky, coarse texture to the landscape with their showy white to pink flowers and interesting green fruit. They are found naturally in dry sandy soils, but they can adapt to other conditions. They colonize by sending up new shoots from the main plant. The tips of their leaves are quite sharp and can be hazardous to people and animals. Butterfly host plants. Maintenance: If they overgrow their spots, cut off the taller shoots at the ground level.

## Palms

While palms and palmettos act as trees and shrubs in the landscape, they do not produce wood and their roots do not expand. This makes palms a good choice to plant near sidewalks and other infrastructure that might eventually be broken up by real trees with expanding roots. Also, because palms do not produce wood, they cannot heal gouges in their trunks, so place them where they will not be injured by power tools or vehicles.

**Bluestem palmetto** (*Sabal minor*) (zones 7–10). This large, tough palmetto shrub has leaves or fronds up to 4 feet (1.2 m) across. It is tolerant of salt spray and inundation by brackish or saltwater. If it grows in standing water, it can grow a trunk, but even in drier areas it can still grow to 9 feet (2.7 m) tall. Showy flowers and berries on tall stalks. Butterfly host plant. Rain garden plant. Maintenance: It does not require any ongoing care.

**Cabbage palm** (*Sabal palmetto*) (zones 8–11). This is Florida's (and South Carolina's) state tree, even though it is not a true tree botanically because it does not produce wood. It prefers moist, rich, somewhat alkaline soil but can adapt to other habitats. It is a medium to large tree and grows slowly; it will be shrub-like for ten years or more before it develops a trunk. It tolerates salt spray and occasional inundation by brackish or saltwater. The six-foot long inflorescence bears fragrant off-white flowers that attract a wide variety of pollinators. The berries are eaten by birds and other wildlife. Butterfly host plant. Maintenance: It is normally sold as a field-grown tree with almost all its fronds trimmed away. At no other time should the fronds be trimmed back as severely, and it should never be topped because that will kill it. Since palms cannot heal gouges in their trunks, do not grow turf right next to the trunk where lawn-care machinery can do damage. Self pruning, so the "hurricane trimming" is not necessary and if done on a regular basis, it will damage the tree.

**Paurotis palm** (*Acoelorrhaphe wrightii*) (zones 10–11). A clumping palm that is tolerant of inundation with brackish water and tolerant of salty wind and may get some salt spray. Maintenance: No particular care is needed.

**Royal palm** (*Roystonea regia*) (zones 10–11). This tall palm is widely planted in South Florida and with good reason. It occurs naturally in moist to wet sites but adapts well to more upland sites once established. Somewhat tolerant of salt spray. Good for large rain gardens. Maintenance: This palm is self-trimming, and its huge fronds, weighing 40 pounds (18.1 kg) or more, will fall when they are ready, so be mindful of that when locating them.

**Saw palmetto** (*Serenoa repens*) (zones 8–11). A long-lived palmetto shrub that can adapt to many habitats. It is tolerant of salt spray and inundation by brackish or saltwater. Most plants have green leaves, but plants with silver leaves are found on Florida's east coast. Berries are eaten by birds and other wildlife. The trunk is recumbent. Highly flammable. Butterfly host plant. Maintenance: Older fronds will die but will eventually be absorbed into the landscape. Some people trim the old fronds, but they can be laid under the shrub to serve as mulch.

## Forbs

Forbs are herbaceous flowering plants that are not grass-like. This is a huge and varied group.

**Asters** (*Symphyotrichum* spp.) (zones 3–11). Asters occur in many different habitats, and in a meadow or butterfly garden, it is likely that at least one of the two dozen Florida native asters will volunteer there. Some are annual and others are perennial, but most will bloom from late summer into early winter and will attract many late pollinators. Scaleleaf aster (*S. adnatum*), climbing aster (*S. carolinianum*), and eastern silver aster (*S. concolor*), are tolerant of salt spray, but not inundation by brackish or saltwater. There are three saltmarsh asters (*S. squamatum*), (*S. subulatum*), and (*S. tenuifolium*) that tolerate both salt spray and inundation by brackish or saltwater. Butterfly host plant. Maintenance: Will withstand annual mowing in a meadow or manual cutting back, but it is not necessary for the plant. Will reseed and spread via rhizomes.

**Blackeyed Susans** (*Rudbeckia* spp.) (zones 2–10). The standard blackeyed Susan (*R. hirta*) is the most widespread and most widely available of the nine *Rudbeckia* species native to Florida. Most species are not salt tolerant, but orange coneflower (*R. fulgida*) and the cutleaf coneflower (*R. laciniata*) are tolerant of salt spray but not of inundation by brackish or saltwater. All species are ruderal plants and occur naturally in sunny meadows, roadside ditches, and other disturbed habitats. They attract pollinators and seed-eating birds. Most blackeyed Susans are short-lived perennials, but they usually self-sow if the conditions are right. Butterfly host plants. Maintenance: While they can adjust to drier conditions, they will do better with some irrigation during droughts.

**Blanket flower** (*Gaillardia pulchella*) (zones 6–11). Blanket flower is a colorful, short-lived perennial known for its bright red and yellow daisy-like flowers. It thrives in well-drained sandy soils and is a popular choice for coastal landscaping due to its drought tolerance and vibrant blooms. It self-seeds. (While this plant is no longer considered to be a Florida native, it is attractive and easy to grow.) Lanceleaf blanketflower (*G. aestivalis*) is native to Central and North Florida and could be an alternative, but it is not salt tolerant.

**Coastal mock vervain** (*Glandularia maritima*) (zones 9–11). This short-lived, evergreen perennial occurs naturally in coastal alkaline sandy soil but is adaptable. It tolerates salt spray but not inundation by brackish or saltwater. Its showy purple flowers attract pollinators. Butterfly larval host. Endangered in Florida. Maintenance: No requirements once established.

**Glassworts** (*Salicornia* spp.). Glassworts are succulent annuals that grow in salt marshes and on muddy shores. They have jointed, green stems and are notable for their lack of distinct leaves. They are often used in saline agriculture and for soil bioengineering due to their salt tolerance. Maintenance: No requirements once established.

**Goldenrods** (*Solidago* spp.) (zones 3–11). The more than twenty species of goldenrod native to Florida are mostly fall-blooming perennials that work well for meadows; they spread by bunching or via rhizomes to form wide-spreading colonies. Most are not salt tolerant, but wand goldenrod (*S. stricta*) and seaside goldenrod (*S. sempervirens*) are tolerant of salt spray and inundation by brackish or saltwater. Some are drought tolerant, while others do better under irrigation. Most have some allelopathic tendency to keep other plants at bay. (*Note:* Goldenrods do not cause hay fever—they have big sticky pollen grains that are carried away by their pollinators, so the pollen is too heavy to be carried in the air. The wind pollinated ragweed (*Ambrosia* spp.) blooms at the same time, but their flowers are not showy, so people do not notice them and blame the goldenrods.) Two other genera are also called goldenrods, *Euthamia* and *Bigelowia;* they are rarely found for sale, but they may volunteer in meadow areas. Butterfly host plants. Maintenance: Throughout most of their range, the stalks die at the end of the season. They can tolerate an annual mowing in the winter. Most spread by rhizomes.

**Hibiscus** (*Hibiscus* spp.) (zones 7–11). There are seven hibiscus species native to Florida and most are adapted to moist or wet habitats, but most are not salt tolerant. Swamp rosemallow (*H. moscheutos*) is tolerant of salt spray and inundation by brackish or saltwater; and sleepy hibiscus (*H. furcellatus*) is tolerant of salt spray but not of inundation by brackish or saltwater. Most of the native hibiscus species have showy flowers in spring and summer, and most die back in the winter in areas where there are frosts. Butterfly host plant. Maintenance: No particular maintenance is needed.

**Sages** (*Salvia* spp.) (zones 7–11). There are seven sages native to Florida and most are annuals or short-lived perennials. Most are not salt tolerant; but tropical sage (*S. coccinea*) is tolerant of salt spray but not of inundation by brackish or saltwater. As members of the mint family, they all have beautiful tubular flowers that attract butterflies, bees, and hummingbirds. They grow in full sun or partial shade in poor neutral to slightly alkaline soils but adapt to many habitats. Butterfly larval host plants; seeds attract birds. Maintenance: Most reseed freely. Pull or transplant seedlings that sprout in inconvenient places.

**Scorpiontail** (*Heliotropium angiospermum*) (zones 9–11). It is an annual or short-lived perennial that grows to about 2 feet (0.6 m). It is native to coastal regions in Central and South Florida and is noted for its showy white flowers. It is somewhat tolerant of salty storm surges and salt spray. There are three other related forbs that also have tolerance to salt spray. Pineland heliotrope (*Euploca polyphylla,* formerly *H. polyphylla*), which is native to most of peninsular Florida; seaside heliotrope (*H. curassavicum*), which is native to coastal regions of peninsular Florida and is also tolerant of occasional inundation by saltwater; and Key West heliotrope (*Euploca fruticosa*), which is native to the Florida Keys and has some tolerance of occasional inundation by brackish water. Maintenance: Since they are annuals in more northern regions of Florida, they may need to be replanted every few years to keep the populations viable.

**Sea purslane** (*Sesuvium portulacastrum*) (zones 6–11). This is a fleshy, drought-tolerant ground cover that occurs naturally on dunes and alkaline soils. It is tolerant of salt spray and inundation by brackish or saltwater. The small lavender flower is attractive, but the plant's hardiness is its best asset. Butterfly larval host. Maintenance: No particular maintenance is necessary if given room to spread; otherwise trimming back from sidewalks may be necessary.

**Spiderlilies** (*Hymenocallis* spp.) (zones 6–11). There are more than a dozen species of spiderlily that are native to Florida. They are large perennials with large bulbs with showy white flowers and most are adapted to moist or wet freshwater habitats. The mangrove spiderlily (*H. latifolia*) is the exception since it tolerates salt spray and frequent inundation by brackish or saltwater. Its natural range is coastal areas in Central and South Florida. Maintenance: The large leaves may die back in the winter, which may look messy, but trimming them back is not necessary or recommended.

**Spotted beebalm,** or dotted horsemint (*Monarda punctata*) (zones 3–10). This tough bunching perennial has an unusual flower structure that is attractive to humans but particularly so to pollinators. It is tolerant of salt spray but not of inundation by brackish or saltwater. When it occurs on the backsides of sand dunes, those plants are shorter, and their bracts are pinker than those grown in moist, salt-free habitats. This is a wonderful addition to pollinator gardens or meadows. It may be more short-lived in more southern regions. It has the same oil as thyme and oregano, so it can be added as a crop to vegetable gardens. Maintenance: It has a rangy habit and stems can lean over as it grows. If it is growing along an edge, prune the stems near the edge back somewhat in early summer. Old stalks turn brown and may need to be cut back if growing in a managed area of the landscape; but deposit those trimmed stems in a brush pile so birds can find those seeds and so overwintering bugs can complete their life cycles. Reseeds to some degree if the conditions are right.

**Standing cypress** (*Ipomopsis rubra*) (zones 5–9). This biennial has bright red flowers attractive to hummingbirds and butterflies. It is great for dry meadows and is tolerant of salt spray and occasional inundation by brackish water. Reseeds. Maintenance: Can tolerate annual winter mowing at a high level.

**Sunflowers** (*Helianthus* spp.) (zones 4–11). At least fourteen species of sunflowers are native to Florida. Some are annuals, while others are perennials. Sunflowers occur in a wide variety of habitats, and most are not salt tolerant. The three variants of the ground-hugging dune sunflower (*H. debilis*) are tolerant of salt spray and occasional inundation by brackish or saltwater. The swamp sunflower (*H. angustifolius*) is tolerant of salt spray but not of inundation by saltwater. Sunflower species attract pollinators and seed-eating birds. Butterfly host plants. Maintenance: Most sunflowers are allelopathic, which may have a negative effect on plants around them, so do not use them as companion plants; but sunflower trimmings can be used as a path mulch to suppress weeds.

**Turkey tangle fogfruit** (*Phyla nodiflora*) (zones 6–11). This widespread plant grows coast to coast and in all parts of Florida and is tolerant of salt spray but not of inundation by brackish or saltwater. It serves as a ground cover and as a turfgrass substitute. It attracts pollinators. Butterfly host plant. Maintenance: No particular maintenance needed, but if it grows beyond its bounds, trim it back.

**Virginia saltmarsh mallow** (*Kosteletzkya pentacarpos*) (zones 4–11). This beautiful mallow with showy pink flowers occurs naturally in fresh and brackish water marshes in the coastal regions of Florida except for the Keys. It is tolerant of salt spray and of inundation by brackish water and occasional saltwater incursions. It can grow to 4 or 5 feet (1.2 to 1.5m) tall and normally dies back in the winter. It is not drought tolerant and needs reliably wet or moist soil. It attracts pollinators. Maintenance: No particular maintenance needed.

## Ferns

**Leather ferns** (*Acrostichum* spp.) (zones 9–11). There are two species native to Florida and both of these robust evergreen ferns require wet or moist habitats with rich soil. Both are tolerant of salt spray and inundation of brackish water. They have dense rhizomes and can be used to control erosion on shorelines. Giant leather fern (*A. danaeifolium*) has a wider range and can grow from 6 to 10 feet (1.8 to 3 m) tall and is dense enough to use as a hedge. Golden leather fern (*A. aureum*) is smaller and occurs mostly in southwest Florida. Maintenance: No particular ongoing maintenance is necessary as long as they are in moist sites.

## Graminoids

The graminoids include grasses and other grass-like plants including rushes and sedges.

**Beaksedges** (*Rhynchospora* spp.) (zones 7–11). There are more than 50 species of beaksedges native to Florida, and they naturally occur in wet places and along shorelines and roadside ditches. Some are tolerant of salt spray and occasional inundation of brackish water, while others occur only in freshwater habitats. Some species may adapt to somewhat drier habitats if shady. A good choice for damp meadows or rain gardens, or even as a lawn substitute in damp areas. Unlike most sedges, starrush whitetop (*R. colorata*), Florida whitetop (*R. floridensis*), and sand-swamp whitetop (*R. latifolia*) are insect pollinated and produce somewhat showy white bracts that attract the pollinators. Maintenance: They colonize, so provide some space.

**Bluestem grasses** (*Andropogon* spp.) (zones 7–11). There are about 20 species of these tough bunching grasses that are native to Florida. They can grow in acidic, poor soil. Because of their bluish leaves and showy inflorescences, they are a great choice for a meadow or to fill in between newly planted native trees while waiting for them to grow. Most are not salt tolerant, but hairy bluestem (*A. longiberbis*) and splitbeard bluestem (*A. ternarius*) are tolerant of salt spray but not of inundation by brackish or saltwater; and big bluestem (*A. gerardii*) is tolerant of salt spray and occasional inundation by brackish water. Butterfly host plants. Maintenance: In meadows and fire-prone habitats, they can be cut back toward the end of winter.

**Bulrushes** (zones 3–9). There are more than a dozen bulrushes native to Florida and all are emergent plants that are important for holding shorelines in freshwater or brackish water habits. The flower heads can be showy or hardly noticeable. The size varies, but they all spread. Saltmarsh bulrush (*Bolboschoenus robustus*) and threesquare bulrush (*Schoenoplectus pungens*) are the most salt tolerant. Some others grow in slightly brackish water, and others still will only survive in freshwater. Purchase from reputable sources that will know the various salt tolerances. Butterfly host plants. Maintenance: No particular maintenance except to make sure they stay wet.

**Cordgrasses** (*Spartina* spp.) (zones 3–11). The cordgrasses are the major components of salt marshes all over the east and Gulf coasts. Five species are native to Florida—all are deep-rooted, tough, and salt tolerant. Smooth cordgrass (*S. alterniflora*) is the most used grass for living shorelines. Saltmeadow cordgrass (*S. patens*) is most often used in high marsh areas that are only periodically inundated. Cordgrasses are highly flammable. (*Note:* Some botanists place the *Spartina* species in the *Sporobolus* genus; if this categorization is accepted, smooth cordgrass would be *Sporobolus alterniflorus,* and saltmeadow cordgrass would be *Sporobolus pumilus.*) Butterfly host plant. Maintenance: No particular maintenance is required if planted in appropriate sites.

**Dropseeds** (*Sporobolus* spp.) (zones 7–11). There are several species of dropseed grasses native to Florida, but one that is noted for its salt tolerance is seashore dropseed (*S. virginicus*), which occurs in salt marshes and is tolerant of both salt spray and inundation by saltwater. Maintenance: No particular maintenance is required if planted in appropriate sites.

**Fakahatchee grasses,** or gamagrasses (*Tripsacum* spp.) (zones 5–11). These are tough, bunching grasses that are somewhat tolerant of salt spray but tolerant of only occasional inundation by brackish water. There are two species native to Florida: the larger and more widespread eastern gamagrass (*T. dactyloides*) and the dwarf Fakahatchee grass (*T. floridanum*), which occurs in South Florida. They both add interest to the landscape and offer good habitat value. Butterfly host plants. Maintenance: It requires no particular maintenance after establishment, but it can be divided if it outgrows its space.

**Lovegrasses** (*Eragrostis* spp.) (zones 3–11). A large number of lovegrass species occur in Florida, but many are not native. Most do best in loamy, slightly moist soils. They are bunching grasses with stems that often arch over giving them a fountain or weeping form—use them as edge plants to highlight this form. The two species most often available from native nurseries: Elliot's lovegrass (*E. elliottii*), a pale green grass that is tolerant of salt spray but not of inundation by brackish or saltwater; and purple lovegrass (*E. spectabilis*), a bunching grass with purple fruits that is not tolerant of salt spray or inundation by brackish or saltwater. Butterfly host plants. Maintenance: They could be cut back in late winter, but it is not necessary unless they are in fire-prone habitats.

**Muhly grass** (*Muhlenbergia capillaris*) (zones 4–11). This easy-to-grow grass is a tough, drought tolerant (once established) but beautiful bunching grass with pink flower heads and a favorite for native landscapes. It is tolerant of salt spray and of occasional inundation by brackish or saltwater. Small seed-eating birds will consume the seeds. Two other Florida species, gulfhairawn muhly (*M. sericea*) and cutover muhly (*Brachyelytrum erectum*) are sometimes sold as *M. capillaris,* but any of the three is a good addition to any landscape. Butterfly host plant. Maintenance: Some people cut old stalks late in the winter, but it is not necessary.

**Panic grasses** (*Panicum* spp.) (zones 4–11). There are six species of panic grasses native to Florida. Most are not salt tolerant, but bitter panic grass (*P. amarum*) and switchgrass (*P. virgatum*) occur naturally along brackish shorelines and tolerate salt spray. Most of these grasses are aggressive growers, and in fact, the nonnative torpedo grass (*P. repens*) is a known invasive plant and needs to be removed from shoreline properties. Maintenance: These are spreading grasses, so they may need to be controlled in some situations.

**Paspalum,** or crown grasses (*Paspalum* spp.) (zones 7–11). There are about twenty species that are native to Florida and almost as many that occur in Florida that are not native, including bahia grass (*P. notatum*), which is widely used as roadside turfgrass. Most of the species are apparently not salt tolerant. One notable exception is the seashore paspalum (*P. vaginatum*), which is tolerant of salt spray and inundation of brackish or saltwater. Maintenance: These are spreading grasses, so they may need to be controlled in some situations. Also, most species require damp conditions for their best growth.

**Rushes** (*Juncus* spp.) (zones 3–10). There are more than twenty species of rushes that are native to Florida. They are common along edges of wetlands and bodies of water and their sturdy network of roots makes them useful for holding shorelines. Most grow as emergent plants, but others may also grow in more upland conditions as long as it is damp. Some rushes are not salt tolerant and occur only in freshwater habitats, while others grow in brackish water or saltwater. The two species most often available from native plant nurseries are needle rush or black rush (*J. roemerianus*) and soft rush (*J. effusus* subsp. *solutus*); both can grow as emergent plants in brackish water. The soft rush is so adaptable that it grows as an emergent plant, a shoreline plant, or an upland plant in damp areas. Many of the rushes work well in rain gardens. Maintenance: No particular maintenance is necessary.

**Saltgrass**, shoregrass, spikegrass (*Distichlis* spp.) (zones 9–11). There are two species native to Florida: saltgrass (*D. spicata*) and shoregrass (*D. littoralis*). They grow in tidal salt marshes and are tolerant of salt spray and frequent inundation by brackish or saltwater. Butterfly host plant. Maintenance: Not drought tolerant–they require reliable moisture.

**Sea oats** (*Uniola paniculata*) (zones 6–11). This tough, drought-tolerant, spreading grass has a massive root system that allows it to survive in the ever-shifting sand dune environment. It tolerates salt spray and inundation by brackish or saltwater. It occurs naturally on beaches and sand dunes but grows well in other sunny habitats. It is illegal to dig it up or harvest seed in Florida because of its importance in holding dunes, so purchase from trusted sources. Maintenance: It needs no maintenance once established.

## Vines

**Baybean,** or beach-bean, seaside jackbean (*Canavalia rosea*) (zones 9–11). It is an aggressive, drought-tolerant, mat-forming vine with showy pinkish flowers that makes a great ground cover along coastal dunes and in other dry, sandy areas. It tolerates salt spray and occasional inundation by brackish or saltwater. Often used for dune restoration plantings. Butterfly host plant. Maintenance: In open areas, it needs no maintenance once established, but it will climb trees and shrubs, so it may need to be removed from other plants on a regular basis.

**Butterfly peas** (*Centrosema* spp.) (zones 7–11). There are two species that are native to Florida. The spurred butterfly pea (*C. virginianum*) is more widespread and is somewhat tolerant of salt spray and occurs naturally along brackish shoreline properties. The sand butterfly pea (*C. arenicola*) is endemic to Florida and is threatened. Butterfly host plant. Maintenance: It needs no maintenance once established, but it may need to be trimmed on a regular basis.

**Clustervines** (*Jacquemontia* spp.) (zones 8–11). There are five species native to Florida and these morning glory relatives can serve as a groundcover or can be trellised. The skyblue clustervine (*J. pentanthos*) occurs in South Florida and can tolerate salt spray, but not frequent inundation by brackish or saltwater; while beach clustervine (*J. reclinata*), also native to South Florida, can tolerate inundation by brackish water with occasional flooding by saltwater, and it has a high tolerance of salt spray. Maintenance: Most will need trimming or folding back to contain their growth to a confined area, so they work best in wilder areas.

**Morning glories** (*Ipomoea* spp.) (zones 8–11). These plants are garden favorites for their showy flowers. About half of the two dozen species that occur in Florida are natives. Three species do well out on the sand dunes and are tolerant of salt spray and occasional inundation: beach morning glory (*I. imperati*), moonflower (*I. alba*), and railroad vine (*I. pes-caprae*), but most other species are not particularly salt tolerant and appreciate more fertility and moisture. Butterfly host plants. Maintenance: Most will need trimming to contain their growth to a confined area, so they work best in wilder areas. In addition to growing vigorously and rooting at every node, many morning glories also reseed profusely.

**Passionflower,** or passionvine (*Passiflora* spp.) (zones 6–11). In most of its range these herbaceous vines die back in the winter, but in frost-free zones they are evergreen. A dozen species occur in the wild in Florida, but only half of them are native. Most species are not salt tolerant, but both the purple passionvine (*P. incarnata*) and the corkystem passionvine (*P. suberosa*) are moderately tolerant of salt spray but not of inundation by brackish or saltwater. They attract pollinators and are the larval food for several butterflies, including Florida's state butterfly, the zebra longwing (*Heliconius charithonia*). Butterfly host plant. Maintenance: If new sprouts pop up in inappropriate places, they can be transplanted but do so early in the year for the best success rate.

**Trumpet creeper** (*Campsis radicans*) (zones 4–10). An aggressive, woody vine with showy orange flowers that are pollinated by hummingbirds and butterflies. It tolerates salt spray but not inundation with brackish or saltwater. Maintenance: It has aerial roots which can damage siding, brick, and stucco, so best allowed to climb trees, poles, trellises and similar structures.

## Non-Salt-Tolerant Plant List

These plants should be used only on the upland side of buildings or other barriers that provide some protection from the salt spray or from storm surges or inundation by brackish or saltwater. Do not use them for upland buffers or for properties adjacent to brackish or saltwater sites.

### Trees and Shrubs

**American strawberrybush** (*Euonymus americanus*) (zones 5–9). This tall, deciduous, thicket-forming shrub prefers rich, well-drained soil. It provides interest in the landscape with unusual dark pink and orange fruit. The canes stay green for several years and are mostly upright. Maintenance: No particular management is necessary.

**Azaleas** (*Rhododendron* spp.) (zones 5–9). Azaleas that are native to Florida are deciduous and have showy flowers that are white, yellow-orange, or pink. They prefer rich, well-drained acidic soil and grow best as understory shrubs in partial shade. The fragrant swamp azalea (*R. viscosum*) is the most widespread in Florida, but it is not often found for sale. Maintenance: No particular maintenance is required once established.

**Buttonbush** (*Cephalanthus occidentalis*) (zones 3–10). This tall, deciduous, multi-caned shrub will form colonies or thickets along freshwater shorelines, and it grows well with constant inundation. Its globe-shaped flower heads attract a wide variety of pollinators. It bears more blooms in sunnier habitats, but it will grow well in the shade. Butterfly host plant. Maintenance: If it spreads too much, the new offsets can be transplanted or pulled out. Some trimming may be needed for older shrubs as the canes break off.

**Firebush** (*Hamelia patens*) (zones 9–11). This cold-sensitive shrub becomes woody in frost-free zones but dies back to the ground after frosts. It prefers dry, sandy, alkaline soils but will also grow on limestone and humus-rich hammocks. In frost-free areas, it has a continuous blooming cycle and attracts hummingbirds and butterflies. Songbirds eat its berries. While it is drought tolerant, it grows better with regular irrigation. Butterfly host plant. Maintenance: Readily reseeds. Clear out dead plant material after die-back in its northern ranges.

**Maples** (*Acer* spp.) (zones 3–10). The most common red maple (*A. rubrum*) is a medium to large tree; it is dioecious and deciduous, with good fall and early spring color. It is adaptable but prefers neutral to acidic soil. A good rain garden tree that does well in wet sites, but it also can do well in drier sites. There are four other maple species native to mostly northern parts of Florida. Butterfly host plant. Maintenance: They tend to sucker and have wide-spreading shallow roots and reseed freely. Older trees sometimes develop heartwood rot.

**Oakleaf hydrangea** (*Hydrangea quercifolia*) (zones 7–9). This large deciduous shrub bears beautiful large white to pink flower heads, and its leaves turn reddish in the fall. It is planted in hedgerows for its texture and beauty. It prefers rich, slightly acidic soils, and while it can survive droughts, it will look better with supplemental irrigation during the dry season. Maintenance: No particular maintenance is required.

**Redbud** (*Cercis canadensis*) (zones 4–9). This small deciduous understory tree is in the legume family and can grow in poor soils. Prefers well-drained acidic soils. Pink flowers appear in the spring before its heart-shaped leaves emerge. Not wind tolerant. Butterfly host plant. Maintenance: The bean-like seed pods turn brown in the winter and stay attached to the tree until spring.

**Red mulberry** (*Morus rubra*) (zones 4–11). This easy-to-grow, medium-sized tree bears edible fruit eaten by birds, other wildlife, and humans. Maintenance: The fruit can be messy.

**River birch** (*Betula nigra*) (zones 4–9). This medium to large deciduous tree with shaggy bark provides year-round interest. It prefers acidic soil and adapts to wet or somewhat drier sites. Can be used in a rain garden and is often planted in groups of three or five. Butterfly host plant. Maintenance: This tree has no particular problems.

**Staggerbushes** (*Lyonia* spp.) (zones 8–10). These understory shrubs do best in moist acidic wooded areas. Five species occur in Florida and any of them will add to the habitat with their flowers and fruit. Maintenance: No particular maintenance is necessary.

**Sugarberry** (*Celtis laevigata*) (zones 4–10). A deciduous medium-sized tree, it can grow in either acidic or alkaline soils and also adapts to moist or dry habitats; it is drought tolerant. It grows quickly and provides good habitat for birds. Butterfly host plant. Maintenance: Sugarberry and other hackberries are allelopathic and may have a negative impact on plants growing nearby.

**Viburnums** (*Viburnum* spp.) (zones 5–11). The common name viburnum is applied to both deciduous and semi-evergreen understory shrubs that have flat or global flower heads with white florets that become black or purple berries. They prefer acidic, well-drained soils and full sun to partially shady habitats. Five species are native to Florida, and all would make good additions to hedgerows. They all offer good habitat value. Butterfly host plants. Maintenance: They can adapt to pruning but will probably do better unpruned and will require less maintenance if allowed to grow into their natural shapes.

**Virginia sweetspire** (*Itea virginica*) (zones 5–10). This shrub, which bears fragrant white flowers, may grow from 3 to 8 feet (1 to 2.4 m) tall—or even taller in wet spots. It is native to freshwater wet areas in all but the southernmost counties in Florida. Maintenance: It sends up suckers from the base of the trunk if planted in wet sites, so plan ahead for that. It is not drought tolerant, but other than that, no particular maintenance is needed.

**Willows** (*Salix* spp.), or **Carolina willow** (*S. caroliniana*) (zones 5–11). This deciduous willow is the most common member of the genus in Florida, but a few other native willows could also work in Florida landscapes. They attract pollinators with their early flowers and are a larval host for butterflies. They are naturally found in wet sites and are a good rain garden plant, but they may adapt to drier sites if irrigated well during the establishment period. Willows are not wind tolerant and only somewhat tolerant of salt spray. Maintenance: Most willows send up many suckers from the base of the trunk. The roots aggressively seek water, so do not plant it anywhere near septic or sewer pipes.

**Winged elm** (*Ulmus alata*) (zones 6–9). This elm is a deciduous, fast-growing, medium-sized tree with interesting corky extensions on its young branches. It grows best in moist acidic soils but is adaptable. Offers good habitat value and is a butterfly host plant. Maintenance: May need corrective pruning while young to encourage a single leader for a stronger structure. Any of the elms may be susceptible to Dutch elm disease, but it has not normally been a problem in Florida.

## Forbs

**Arrowheads,** or duck potato (*Sagittaria* spp.) (zones 3–11). These colony-forming aquatic perennials are normally found in wet ditches and pond shorelines. The thirteen species that are native to Florida occur in freshwater. They produce tall flower spikes and bloom over an extended period each year. Butterfly host plant. Maintenance: Can be aggressive, so they are good for shoreline stabilization.

**Blazing stars** (*Liatris* spp.) (zones 6–11). More than a dozen blazing star species are native to Florida. Most are short-lived perennials that die back each winter. They can occur naturally in wet or damp meadows or in sandy uplands. They are attractive to butterflies and other pollinators. Maintenance: No particular maintenance is necessary, but they are short-lived and may need to be replanted every few years to renew the population.

**Blue-eyed grass** (*Sisyrinchium angustifolium*) (zones 3–11). This dainty member of the iris family is a bunching, short-lived perennial with showy pale blue flowers in the spring. Use as a turfgrass substitute in a lawn, in short damp meadows, or at the edges of native areas. Rain garden plant. Maintenance: It tolerates mowing and light foot traffic.

**Blue mistflower** (*Conoclinium coelestinum*) (zones 6–11). This short-lived perennial occurs naturally in moist habitats and dies back in the winter. Its fuzzy blue or lavender flower heads attract many pollinators. Rain garden plant. Butterfly host plant. Maintenance: In moist conditions, it can spread, so give it plenty of room or be ready to divide it.

**Cardinal flower** (*Lobelia cardinalis*) (zones 3–10). This short-lived perennial is a beautiful addition to acidic damp or wet garden areas. Its red tubular flowers attract hummingbirds and butterflies, but it is not usually a good rain garden plant in Florida because of the long dry season. Maintenance: Unless the conditions are just right, it may need to be replanted to retain the population.

**Carolina redroot** (*Lachnanthes caroliana*) (zones 6–11). This is an easy-to-grow and well-behaved perennial as long as the soil is wet and acidic. The white to cream flowers attract a wide variety of pollinators. Dies back in the winter. Yes, the roots are bright red, but do not pull them out to see. Butterfly host plant. Maintenance: No particular maintenance as long as they are wet.

**Iris** (*Iris* spp.) (zone 7–10). There are eight iris species native to Florida. Most prefer moist or wet habitats, and their thick intertwining rhizomes make irises a good choice to stabilize freshwater shorelines. Most have beautiful blue/purple flowers and are usually evergreen, but some may die back in cooler regions. Maintenance: No particular maintenance required.

**Lizard's tail** (*Saururus cernuus*) (zones 4–11). This striking perennial is a colonizing wetland or emergent freshwater plant with showy white flower spikes that droop. It dies back in the winter. It prefers acidic soils and sunny to mostly shady habitats. Maintenance: No particular maintenance as long as it is wet and has room to spread.

**Meadow garlic** (*Allium canadense*) (zones 4–9). This tough perennial garlic, with its flat leaves and unusual inflorescence that includes both bulblets and flowers on stalks, occurs naturally in damp ditches, but it can also do well in drier sites. It dies back in the summer and resprouts in late fall. Rain garden plant. All parts of this plant are edible. Maintenance: No particular maintenance is required, but it will spread when the bulblets drop from the inflorescences. It tolerates mowing.

**Milkweeds** (*Asclepias* spp.) (zones 4–10). There are more than twenty milkweeds native to Florida, and they are beautiful and important additions to a native landscape because they are the exclusive larval food for monarch and queen butterflies. Some species do well in moist habitats, while others work well in dry fields, so research the needs before planting. (*Note:* Scarlet or tropical milkweed (*A. curassavica*) is widely sold, but it is not native to Florida and is harmful to monarch butterflies.) Butterfly host plant. Maintenance: Once established, milkweeds do not require any particular care, and it is adapted to being eaten by caterpillars.

**Mimosas** (*Mimosa* spp.) (zones 8–11). Two species of mimosas are native to Florida and are vining ground covers in the legume family; they can grow well in poor, sandy soils because their roots fix nitrogen. Sunshine mimosa (*M. strigillosa*) makes a beautiful turfgrass substitute with its pink powderpuff flower heads, and it withstands light foot traffic. Sensitive briar (*M. quadrivalvis*) is similar, but it has prickles along the stem. Mimosas attract pollinators. Butterfly host plant. Maintenance: Once established, an annual mowing at the highest setting will probably be needed to keep down the weeds at least for a few years until it forms a thick mat. (*Note:* This is not the invasive tree also known as mimosa [*Albizia julibrissin*].)

**Pickerelweed** (*Pontederia cordata*) (zones 4–11). A clumping, emergent perennial with a showy purple spike of flowers that attracts pollinators. Soil can be acidic to neutral, but it must be wet. Pickerelweed works well by itself or with other freshwater shoreline plants. Various waterfowl eat the seeds. Maintenance: No particular maintenance is required as long as the site stays wet.

**Rain lilies** (*Zephyranthes* spp.) (zones 7–9). These bulbaceous perennials are in the amaryllis family. They occur naturally in wet meadows and ditches but can also grow in drier habitats. Three rain lilies are native to Florida, but the most abundant is the atamasco lily (*Z. atamasca*). Normally they are short-lived, but colonies can last for many years in the right conditions. They tend to bloom in the spring a few days after a heavy rainfall, but by the middle of summer they will die back. The grass-like leaves are not particularly noticeable when they are not blooming. Rain garden plant. Maintenance: It may need to be replanted every few years to maintain a population.

**Rattlesnakemaster** (*Eryngium aquaticum*) (zones 3–10). This interesting member of the carrot family is best used at the edges of fresh bodies of water or wetlands. Its unusual flower heads and basal leaves add good texture to shorelines. Attracts many pollinators. Maintenance: Once it is in place, it does not require any special maintenance as long as it is damp.

**Stokes' aster** (*Stokesia laevis*) (zones 8–9). This low-growing and bunching perennial has showy lavender flower heads that attract pollinators. Good for borders or edges. Does best in sun or partial shade in neutral to acidic soils. Maintenance: Keep other vegetation from flopping over these low-growing plants. Flower stalks turn brown after blooming.

**Tickseeds** (*Coreopsis* spp.) (zones 4–11). There are twelve tickseeds native to Florida—none is salt tolerant. They are meadow wildflowers with mostly yellow flower heads; most are annuals. They are Florida's state wildflower. Most *Coreopsis* species prefer moist, sunny habitats, but some do well in drier conditions. They attract a wide variety of pollinators. Maintenance: Will do well with an annual mowing in the winter. Reseed to maintain their density in long-term meadows.

## Ferns

**Cinnamon fern** (*Osmundastrum cinnamomeum*) (zones 3–11). This is a large clumping fern that makes quite a statement in damp or wet habitats. It will grow well in full sun only if it has reliably wet soil; if the soil dries out for part of the year, it should have full or partial shading. Good for rain gardens that are partially shaded. The spores are borne on fertile fronds that are the color of cinnamon. In Florida, it sends up fertile fronds in both spring and fall, and its sterile fronds appear in the spring and keep growing until late fall when they die back. Maintenance: It is best to keep weeds and vines from overrunning it.

**Royal fern** (*Osmunda spectabilis*) (zones 3–10). This magnificent large clumping fern deserves its common name. Its fronds have fully separated, doubly pinnate leaves, making the fern appear lacy. The spore-producing fertile fronds appear only at the top of its main sterile fronds, so later in the season, after the fertile parts have fallen away, the remaining fronds look cut off. Plant only in habitats that are moist to wet. Fronds die back in the winter. Maintenance: Older specimens may form vertical mounds, which may be invaded by trees or shrubs, so it is best to pull those plants from the mound while they are small. May need to irrigate this fern if planted in drier sites.

**Chain ferns** (*Woodwardia* spp.) (zones 3–11). There are two species native to Florida. Both species form colonies in wooded or partially shaded areas. While they occur naturally in damp sites, they can adapt to drier locations as long as they have some shade. Use them as a ground cover in groves once there is some shade. The netted chain fern (*W. areolata*) is about 1 to 1.5 feet (0.3 to 0.5 m) tall and has separate, stiff, fertile fronds that usually persist through the winter, but the sterile fronds die back where there is frost. The Virginia chain fern (*W. viginica*) is a bit taller, up to 3 feet (1 m) tall, and produces spores on the back of all its fronds. Maintenance: No particular maintenance is required, but it is best to keep vines and other vegetation from overrunning the area where they grow.

## Graminoids

**Maidencane** (*Hymenachne hemitomon*) (zones 6–11). An aggressive grass with deep, persistent rhizomes. While not strictly aquatic, it tolerates standing water and is effective as a shoreline stabilizer for freshwater sites. Maintenance: No particular maintenance is required, but it is best to use it only where its aggressiveness will not be an issue.

## Vines

**Carolina jessamine,** or evening trumpetflower (*Gelsemium sempervirens*) (zones 7–10). This is a fast-growing, evergreen, winter-flowering vine that can be used on trellises and fences or as a ground cover in sunny and partial shade areas. Its showy flowers are also fragrant. It roots readily. Poisonous to livestock. Maintenance: If it is used as a ground cover, it tends to climb vertical obstacles in its path. Even though it is a fast grower, it is relatively easy to control.

**Clematis** (*Clematis* spp.) (zones 4–11). Six species are native to Florida and most are vines, but pine hyacinth (*C. baldwinii*) is not. Some of the clematis species occur near shorelines, but they are not salt tolerant. Maintenance: No particular care is needed after establishment except for trimming it back where neatness counts.

**Coral honeysuckle** (*Lonicera sempervirens*) (zones 4–10). This long-lived vine provides excellent habitat value with its beautiful red tubular flowers that attract hummingbirds and butterflies, and its orange berries are eaten by a wide variety of songbirds. It blooms best if allowed to climb a fence or trellis, but it can also serve as a ground cover. Butterfly host plant. Maintenance: If neatness counts, trim back the vines yearly after the hummingbirds have left for the season.

**Wild allamanda** (*Pentalinon luteum*) (zones 10–11). This long-lived perennial vine is woody. Its showy flowers and glossy foliage make quite a show in full or partial sun. It can grow in poor, sandy soil and is somewhat salt tolerant. Attracts pollinators. Butterfly host plant. Maintenance: No particular care is needed after establishment except for trimming it back where neatness counts.

# Appendix B

# Mulching Materials

## Types of Mulch

A wide variety of organic and inorganic materials may be used for mulching. Local sources for mulching materials are recommended, if possible. If the target habitat is naturally nutrient poor or in shoreline areas, do not use nutrient-rich organic mulches. Also, if the landscape is in a fire-prone area, most organic mulches are flammable and are best used outside of the thirty-foot clear zone around buildings. (For general mulching guidelines, including information on why weed barrier cloth is not recommended for Florida, see chapter 8.)

Each type of mulch has its pros and cons depending on where and how it will be used and its overall sustainability. Here are some considerations for the more commonly used mulches:

### Chipped Tree Trimmings or Arborist Chipped Wood

This is a mix of shredded wood and green leaves—the tree parts that have been run through a chipper—and are available from tree trimming companies. As a mulch, it consists mostly of nicely chopped wood mixed with leaves and some thin branches that slide through the shredder. This mulch will last for a year or more as a garden or pathway mulch. It is not recommended for areas that are flooded on a regular basis because the wood floats. If it is used in areas that are not disturbed, a thick layer of wood chips may form a crust, so in such areas, do not use more than a two-inch (5.1 cm) layer unless this is an attempt to smother and kill the weeds in the area. It will eventually decay and become part of the soil.

Because nothing has been added to it, the look is quite natural, and if leaves fall on it, no removal is necessary—it all blends together. This is one of the most sustainable mulching solutions because the tree trimmers save gas by dumping their chipped tree trimmings in the neighborhood; homeowners save gas by not having to drive somewhere to pick it up or not having someone to deliver mulch; the landfills are not used; and the homeowners save money. This mulch is also sustainable because the production of bagged mulch, no matter what type, uses energy in its packaging and delivery.

## Compost

Unfinished compost, where the shapes of some of the original ingredients are visible, provides a soil-enriching mulch. In this state, the microbes are still working to decompose the materials and may extract available nitrogen from the soil to live. This prevents weed seeds from germinating. Compost moderates the temperature and moisture of the soil below because it is organic and dark in color. Since its texture is similar to that of the soil, it provides less of a physical barrier than coarser mulches. Finished or completed compost is probably better used in the landscape as a top dressing around plantings. Because it does not last long before it becomes part of the soil, it is only somewhat sustainable. It is not recommended for areas that are below the mean higher high-water lines (MHHW) because this nutrient-rich material could wash into the water or wetland areas.

Since it is filled with soil microbes, finished or unfinished compost can be used to help restore soil ecosystems in sterile subsoil. In shoreline areas, use with caution so that no compost will blow or wash into the water. (See sidebar "Soil Ecosystems" in chapter 9.)

## Gravel, Coarse Sand, or Crushed Shells

Gravel, coarse sand, or crushed shells are inorganic materials that are often recommended for areas that are below the MHHW line because they tend to stay in place when flooded and do not add to the nutrient levels in the water if they are washed over by water. These materials have several negatives as a mulch, though. In Florida these materials do not do much to deter weeds from growing, so weeding will become an added chore for its maintenance. Rocks, big and small, retain and reflect the heat long after the sun has set; and a layer of rocks can turn a sunny spot into an oven. On top of that, it is difficult to keep gravel clear of leaves and other debris where more weeds could then get started. Plus, once it is laid in, it will be impossible to remove it all from the soil. If this is the choice for a shoreline mulch, find a source for local materials such as crushed coquina rock, oyster shells, or sand instead of imported materials such as volcanic gravel.

## Hay or Straw

Hay consists of grasses and forbs that have been allowed to go to seed before mowing and gathering. Straw generally does not include the tops or seed heads, but just the stalks, so it contains fewer weed seeds and would be preferable over hay for mulch in most cases, except for maybe in a meadow, where the variety of seed could add to the diversity. It is effective as a mulch, but do not pile it too thickly or it can become anaerobic and start to smell sour. In a native bed, the look will not be natural, but if that is important, it can be covered with a thin layer of a more natural-looking mulch such as pine needles, leaves, or wood chips. They generally are sustainable since they are made from waste organic materials. When newly applied, hay or straw could float away in areas that are inundated.

## Leaves or Shredded Leaves

Of course, leaves are the most natural of mulches. As leaves fall from trees and shrubs, they mulch the surrounding areas. It is best to let the leaves remain under the trees or shrubs from which they fell—they supply just the right mix of nutrients. When leaves fall on lawn or hardscape areas, rake them into the surrounding beds or save extra leaves in a large pile in an out-of-the-way location in the landscape. If the leaves are large or leathery, some people shred or compost them before using them as mulch. Shredded leaves do not last as long, but they may stay in place better than whole leaves. Leaves or shredded leaves are not recommended for areas below the MHHW line, because they can wash away and add too many nutrients to the water. Some people collect their neighbors' bags of leaves in the fall as they put them on the curb for pickup. This is a good way to build up a large supply to use throughout the year, especially if the trees are still small. Highly sustainable.

## Peat Moss

Peat moss, which is mined mostly from *Sphagnum* peat bogs in Canada and northern Eurasia, is often used in gardens for mulch and more. The peat formed in bogs builds up slowly (at about one millimeter per year) and because it is formed so slowly, it can never be mined sustainably. In addition to not being a sustainable product, peat moss is so acidic that it kills soil microbes and offers almost no nutrients for plants. It is not recommended for any landscape for any purpose. To accomplish many of the same purposes, such as peat pots and hanging basket liners, use those made from coconut coir or cow manure instead.

## Pine Bark or Other Bark

Bark's purpose is to protect the tree from injury and to prevent water loss, so it contains waxy materials to repel water. Use only a two or three-inch (7.6 cm) layer of bark. As a mulch, bark products usually last longer than other organic materials because of the water-repellent chemicals. Bark nuggets may be easier to handle, and they tend to stay in place better than the larger bark chunks unless there is a lot of water flowing through the landscape. Bark mulches do not form a crust. Do not use bark for mulching rain gardens, bioswales, or near shorelines because it floats. Bark is usually a by-product from lumber mills and is generally a sustainable product. This is the least flammable of the organic mulch types.

## Pine Needles or Pine Straw

Pine needles make an attractive durable mulch that does not readily float or form a water-resistant crust. Use it for paths, in garden beds, or around trees and shrubs. It has been shown that while green pine needles are quite acidic, the soil under a pine needle mulch becomes only slightly more acidic. Pine needles make such a long-lasting and good-looking mulch that they are harvested from tree farms and sold as pine straw, but they are easily gathered from streets and neighborhoods if there are pines in the area. A waxy or resinous coating on the needles makes pines more drought resistant. It also makes pine needle mulch resist decomposition more than other materials. It is sustainable, especially if gathered locally. Be aware of invasive weeds included with the purchased bales—Japanese and Old World climbing ferns (*Lygodium japonicum* and *L. microphyllum*) have been found in Florida's pine straw bales.

## Plastic Sheeting

Plastic sheeting is not recommended for native or natural landscapes. When it is used as a mulch, plastic is impenetrable by even the most vigorous of Florida's weeds, but it also disrupts the soil's ecosystem. In many commercial agricultural operations, plastic sheeting is stapled to the soil between vegetable rows. In order to have it work well, commercial operations use drip irrigation to water the plants because the plastic does not allow water to penetrate. In addition, plastic heats the soil, and in fact, it is recommended for "solarizing" the soil to kill plants, weed seeds, nematodes, and other microbes. It is not sustainable and not recommended.

## Sawdust

A by-product from lumber mills, the small particles pack together and often create anaerobic conditions if held in a pile for too long. As a mulch, sawdust reduces nitrogen from surrounding soils more than other mulches. This means that it is a good weed suppressant, but not necessarily good to use around new seedlings or newly planted natives. It will eventually enrich the soil, so it is not recommended for shoreline projects. Do not use sawdust from chemically treated wood. Sawdust is generally sustainable since it is a by-product.

## Shredded Wood

Shredded *Melaleuca* is a sustainable choice because using this mulch may help with the efforts of reducing these invasive trees from Florida's ecosystems. Shredded *Eucalyptus* is also a reasonable choice, but because it is farmed, using this mulch does not necessarily help the effort to clear it from wild areas. Cypress is probably Florida's most popular mulching material—dyed in several colors. It used to be a by-product from lumber mills, but now whole cypress forests are being ripped out to accommodate the demand. So, while this may be a good mulching product, it is not sustainable because the high demand has outstripped the supply. It is not recommended for use in a shoreline landscape because it can float when flooded after it has first been laid in; but after a time, it tends to be absorbed into the soil and enriches it.

## Shredded Rubber or Shredded Tires

Do not use shredded rubber or shredded tires. Some studies have shown that rubber leaches potentially harmful and persistent chemicals into the soil. Rubber mulch is highly flammable, and it does not add nutrients to or improve the structure of the soil beneath it. Not sustainable.

## Spanish Moss

Spanish moss (*Tillandsia usneoides*) is a flowering plant in the pineapple family (Bromeliaceae) that is native to the southeastern states and to some of the Caribbean, so it is not Spanish and it is not a moss. (Mosses are nonvascular, nonflowering, spore-producing plants.) It is an epiphyte, or air plant, found in damp areas near waterways or swamps because it absorbs its moisture and nutrients from the air, so while it hangs from trees, it is cleaning the air. It is not a parasite and does not rob the host tree of water or nutrients.

It does not need to be removed from trees, but if there is an abundance of fallen Spanish moss, it serves as an effective and durable mulch. It is sustainable since there is no manufacturing or packaging involved.

# Appendix C

# Permitting Language

## Federal Permit Instrument Language

Source link: https://www.swt.usace.army.mil/Portals/41/docs/missions/regulatory/2021%20NWP/NWP-54.pdf.

## Nationwide Permit 54—Living Shorelines

Effective Date: February 25, 2022;
Expiration Date: March 14, 2026

(NWP Final Notice, 86 FR 73522)
**Nationwide Permit 54—Living Shorelines.** Structures and work in navigable waters of the United States and discharges of dredged or fill material into waters of the United States for the construction and maintenance of living shorelines to stabilize banks and shores in coastal waters, which includes the Great Lakes, along shores with small fetch and gentle slopes that are subject to low- to mid-energy waves. A living shoreline has a footprint that is made up mostly of native material. It incorporates vegetation or other living, natural "soft" elements alone or in combination with some type of harder shoreline structure (e.g., oyster or mussel reefs or rock sills) for added protection and stability. Living shorelines should maintain the natural continuity of the land-water interface, and retain or enhance shoreline ecological processes. Living shorelines must have a substantial biological

component, either tidal or lacustrine fringe wetlands or oyster or mussel reef structures. The following conditions must be met:

(a) The structures and fill area, including sand fills, sills, breakwaters, or reefs, cannot extend into the waterbody more than 30 feet from the mean low water line in tidal waters or the ordinary high water mark in the Great Lakes, unless the district engineer waives this criterion by making a written determination concluding that the activity will result in no more than minimal adverse environmental effects;
(b) The activity is no more than 500 feet in length along the bank, unless the district engineer waives this criterion by making a written determination concluding that the activity will result in no more than minimal adverse environmental effects;
(c) Coir logs, coir mats, stone, native oyster shell, native wood debris, and other structural materials must be adequately anchored, of sufficient weight, or installed in a manner that prevents relocation in most wave action or water flow conditions, except for extremely severe storms;
(d) For living shorelines consisting of tidal or lacustrine fringe wetlands, native plants appropriate for current site conditions, including salinity and elevation, must be used if the site is planted by the permittee;
(e) Discharges of dredged or fill material into waters of the United States, and oyster or mussel reef structures in navigable waters, must be the minimum necessary for the establishment and maintenance of the living shoreline;
(f) If sills, breakwaters, or other structures must be constructed to protect fringe wetlands for the living shoreline, those structures must be the minimum size necessary to protect those fringe wetlands;
(g) The activity must be designed, constructed, and maintained so that it has no more than minimal adverse effects on water movement between the waterbody and the shore and the movement of aquatic organisms between the waterbody and the shore; and
(h) The living shoreline must be properly maintained, which may require periodic repair of sills, breakwaters, or reefs, or replacing sand fills after severe storms or erosion events. Vegetation may be replanted to maintain the living shoreline. This NWP authorizes those maintenance and repair activities, including any minor deviations necessary to address changing environmental conditions.

This NWP does not authorize beach nourishment or land reclamation activities.

*Notification:* The permittee must submit a pre-construction notification to the district engineer prior to commencing the construction of the living shoreline. (See general condition 32.) The pre-construction notification must include a delineation of special aquatic sites (see paragraph (b)(4) of general condition 32). Pre-construction notification is not required for maintenance and repair activities for living shorelines unless required by applicable NWP general conditions or regional conditions. (Authorities: Sections 10 and 404)

*Note:* In waters outside of coastal waters, nature-based bank stabilization techniques, such as bioengineering and vegetative stabilization, may be authorized by NWP 13.

## State Permit Living Shoreline Exemption Language

Excerpted from Florida Administrative Code: Florida Small-Scale Living Shoreline Exemption (Florida Administrative Code 62-330.051(12)(e). Source link: https://www.flrules.org/gateway/ruleno.asp?id=62-330.051.

### 62-330.051(12)(e) Exempt Activities.

The activities meeting the limitations and restrictions below are exempt from permitting. However, if located in, on, or over state-owned submerged lands, they are subject to a separate authorization under Chapters 253 and 258, F.S., as applicable.

. . .

(12) Construction, Replacement, Restoration, Enhancement, and Repair of Seawall, Riprap, and Other Shoreline Stabilization—

. . .

(e) Restoration of an eroding shoreline with native wetland vegetative enhancement plantings, provided:

1. The length of shoreline is 500 linear feet or less;
2. Plantings are native wetland plants appropriate for the site obtained from commercially-grown stock;
3. Plantings extend no farther than 10 feet waterward of the approximate mean high water line (MHWL) or ordinary high water line (OHWL);
4. All invasive and exotic vegetative species along the shoreline is removed in conjunction with the planting to the extent practicable;

5. Biodegradable natural fiber logs or mats that are secured in place, such as with the use of wooden stakes, may be used if necessary to support the vegetative plantings; and

6. No fill is placed other than that needed to support the vegetative plantings, except that a "breakwater" is authorized to be installed concurrent with the planting if permanent wave attenuation is required to maintain the shoreline vegetation, provided:

a. The waterward toe of the breakwater extends no more than 10 feet waterward of the approximate MHWL or OHWL, with a top height of no more than the mean or ordinary high water elevation;

b. The breakwater is composed predominantly of natural oyster shell cultch (clean and fossilized oyster shell) or other stable, non-degradable materials such as oyster reef, reef balls, boulders, clean concrete rubble, rip rap, rock sills, or triangular concrete forms. Oyster shell cultch, if used, shall be enclosed in mesh bags having openings of no more than three inches (7.6 cm), or securely fixed to matting prior to placement in the water. Oyster bags and mats must be anchored to prevent movement of shell from the project area;

c. The breakwater shall not be placed over, or within three feet (in any direction) of any submerged grassbed or existing emergent marsh vegetation;

d. The breakwater shall be placed in units so that there is at least one opening measuring at least five feet in width located every 75 linear feet along the breakwater, with a minimum of one opening, to allow the flow of water and the passage of fish and aquatic wildlife;

e. All equipment used during construction shall be operated from, and be stored in uplands; and

f. All work is conducted in compliance with subsection 62-330.050(9), F.A.C.

. . .

*Rulemaking Authority 373.026(7), 373.043, 373.4131, 373.4145, 403.805(1) FS. Law Implemented 373.406, 373.4131, 373.4145, 373.415, 403.813(1) FS. History–New 10-1-13, Amended 6-1-18.*

# Glossary

**ACCRETION:** The opposite of erosion. Accretion occurs when there is a net sediment capture on a shoreline.

**ACIDITY:** Acidity is measured on the pH scale of 1 to 14 (pH is short for the potential of hydrogen). The soil's acidity determines the uptake rates of various nutrients. A pH of 7 is neutral. Substances measuring below 7 are acidic, while those that measure above 7 are alkaline, or basic. Most of Florida's soils are moderately acidic to neutral, but in much of South Florida, the underlying limestone creates an alkaline environment.

**ALLELOPATHY:** Allopathic plants produce biochemicals that harm or destroy nearby plants by inhibiting seed sprouting, root development, or nutrient uptake. Sunflowers and goldenrods are examples of allopathic plants that gain an advantage in their local environment.

**ARMY CORPS OF ENGINEERS:** *See* US Army Corps of Engineers.

**ASSUMED WATERS:** State-assumed waters include all Waters of the United States (WOTUS) where section 404 permitting authority is not retained by the US Army Corps of Engineers (USACE).

**BASEFLOW:** Also called drought flow, groundwater recession flow, low flow, low-water flow, low-water discharge, and sustained or fair-weather runoff. It is the portion of the streamflow that is sustained between precipitation events and fed to streams.

**BENTHIC:** An adjective describing anything that occurs on or in the bottom of a body of water.

**BENTHIC SURVEY:** A survey that characterizes benthic habitats within the survey area that may possibly be affected by the proposed project. These surveys are often required in living shoreline permits.

**BIODEPOSITS:** Oysters are filter feeders. Some particles, such as microalgae, are sent into the oyster's mouth where they are digested, and waste is excreted as feces. However, particles, such as sediment, are usually rejected by an oyster and deposited as "pseudofeces" just outside the shell. They are not digested. Biodeposits are a combination of both feces and pseudofeces.

**BIODIVERSITY:** The biological diversity in an environment as indicated by the numbers of species and genetic variability of plants, animals, fungi, and various microorganisms. All living organisms, their interactions, and their habitats come under the umbrella of biodiversity, which measures how much variability there is among them all.

**BREAKWATER:** An offshore structure built to provide a barrier to or protection from waves. Breakwaters are designed to reduce incoming wave energy, resulting in calmer water between the shoreline and the structure. In Florida, breakwaters built as part of small-scale living shorelines are exempt from permitting only if they are built no higher than the mean high water (MHW) line.

**CERTIFYING AGENCY:** This is the agency or entity designated to certify compliance with applicable water quality standards. Certifying agencies include the Florida Department of Environmental Protection (FDEP), the Florida Water Management Districts (WMD), the Miccosukee Tribes, and the Seminole Tribes.

**CLEAN OYSTER SHELL:** Oyster shells used in living shorelines must be clean of disease and pests to be a safe and allowable material for construction. There are several sources of clean oyster shell, including recycled shell and fossilized shell, from both nonprofit and for-profit organizations. Recycled shells that come from seafood restaurants must be cured outdoors in a sunny area in piles that are routinely turned for at least six months. This prevents the spread of oyster pests and diseases. Permitting agencies will usually request information about the specific source of the shell when reviewing a permit application.

**COASTAL SQUEEZE:** The loss of intertidal habitat caused by rising sea levels in front of an artificially fixed shoreline such as bulkheads.

**COASTAL ZONE MANAGEMENT (CZM):** Section 307 of the Coastal Zone Management Act (CZMA) requires that federal actions that have reasonably foreseeable effects on any coastal use or natural resource of the coastal zone be consistent with the enforceable policies of a coastal zone management agency's federally approved coastal management program.

**COIR/FIBER LOGS:** Coir or fiber logs are a cloth tube or sock filled with meshed natural fibers, such as coir (coconut fiber), to create the shape of a tree log. Coir logs are used to provide temporary protection at the toe of a planting zone to retain sediment and reduce flow; they help to retain the soil while new plantings establish good root systems.

**CONCRETE DOME:** Concrete shaped into domes. They are used as one method of oyster reef construction that uses molded, precast concrete forms designed to mimic the attributes of the natural three-dimensional structure of oyster reefs. In some cases, the concrete mixture is fortified with an additive to increase oyster recruitment. There are many commercially available forms, such as Reef Balls.

**CROSS-SECTION DRAWING:** For purposes of this book, a cross-section drawing is a depiction of a living shoreline project in cross-section, including information about the slope of the shoreline, the approximate mean low water (MLW) and mean high water (MHW) locations, and breakwater configurations. It usually provides additional information such as gaps, heights, distance from the MHW line, the spacing of plants, the location of existing structures, and more.

**CRUSHER RUN GRAVEL:** Includes crushed rock with sizes that range from the stated gravel size down to grit. This is not recommended for drywells, but it may be acceptable as rock mulch for high marsh areas.

**CULTCH:** A mass of stones, broken shells, and grit laid down on oyster grounds to furnish points of attachment for the spat. Cultch is often added to living shoreline projects where oyster reefs are to be part of the structure.

**CUTICLE:** Terrestrial plants exude a complex mixture of waxes and cutins that form a coating on their external surfaces called a cuticle. It protects the plant from water loss, UV radiation, and pathogens including fungi and bacteria. To some extent, the cuticle also deters herbivore activity.

**DETRITUS:** Pieces of organic material consisting of decomposing dead plants, animals, their waste, and sediments. If left on top of soil, the detritus will be incorporated into the next layer of soil. Organisms that consume detritus are detritivores.

**DROP ROOTS:** Specialized roots formed by mangroves that arise from the branches and hang down in the air, usually above the mean high tide and aid in the uptake of rainwater. Red mangroves also have prop roots that arise from lower on the trunk and take root on the bottom. *See also* prop roots

**ECOSYSTEM:** A local network of interacting plants, animals, fungi, bacteria, and other organisms in the environment where they live.

**ECOSYSTEM SERVICES:** Ecosystem services are the benefits nature provides to people. These include food and fresh water; regulation of local climate, water, and pollination; soils and nutrient cycling; and aesthetics, cultural heritage values, recreation, and tourism.

**ENVIRONMENTAL RESOURCE PERMIT:** A type of permit required before beginning any construction activity that will affect wetlands, alter surface water flows, or contribute to water pollution in Florida.

**EMERGENT VEGETATION:** Emergent wetland plants are rooted on the bottom of a body of water with stiff or firm stems that stand above the water's surface. That is, these plants "emerge" from the surface of the water. Cattails (*Typha* spp.) are an example. In some cases, emergent plants can tolerate being submerged during a high water event. Lily pads and many marsh plants are also considered emergent plants.

**EROSION:** This is the process of sediment being carried away from an area by water or wind. In coastal environments, the major driving processes of erosion are storms, flooding, wave action, sea level rise, and human activities (i.e., boating, seawall construction, and development). Erosion is a natural process in coastal ecosystems, but it becomes an issue when homes and other hard infrastructure are threatened. *See also* accretion

**ESCARPMENT:** *See* scarp

**ESTUARY:** A broad term used to describe a protected or semiprotected aquatic environment where freshwater meets the sea.

**EVAPOTRANSPIRATION:** The total output of water vapor into the air from the transpiration from plants plus the evaporation from the soil. It usually is measured from a wooded area or any ecosystem.

**EXOTIC:** Exotic species are those that are not native to an ecosystem. Most are not harmful to the local ecosystem. If an exotic species has taken over or damaged local ecosystems, it could be declared an invasive species.

**FACULTATIVE HALOPHYTE:** A plant that can tolerate saltwater habitats but can also live in freshwater. Most species of mangroves fall into this category.

**FETCH:** The area where ocean waves are generated by the wind; the term refers to the length of the fetch area, measured in the direction of the wind.

**FILL:** Any material (sand, concrete, shell, etc.) imported into a landscape or project area and placed on uplands or on submerged lands.

**FLORIDA DEPARTMENT OF ENVIRONMENTAL PROTECTION (FDEP):** There are six FDEP sections: Northwest, Northeast, Central, Southwest, Southeast, and South. A map of the offices is located at https://floridadep.gov/districts.

**GABION:** A wire framework container that provides a structured foundation for shell or rock material. Made of a material similar to chain-link fencing, gabions are filled with shells and ballast; they are then stacked and wired together to construct sturdy, three-dimensional structures.

**GAP SPACING:** Dimensions and spacing recommended or required for openings in proposed structures (e.g., breakwaters). Gaps reduce the risk of entrapment for marine mammals, turtles, and other wildlife.

**GENERAL PERMIT:** A class of permit available from the Florida Department of Environmental Protection (FDEP) that allows expedited permitting for projects that meet certain criteria. If a project meets the criteria for a State Programmatic General Permit (SPGP), the state can issue a general permit that also provides federal permission to construct the project.

**GRAY INFRASTRUCTURE:** For shorelines, gray infrastructure includes seawalls, bulkheads, revetments, riprap, and other built structures.

**HABITAT:** A habitat is the natural home or environment of a plant, animal, or other organism. A complete habitat, consisting of both biotic and abiotic factors, provides the organisms that live there with food, water, shelter, and space to survive.

**HIGH-ENERGY COASTS:** Areas that have strong wave action, where the average rate of erosion over time is usually larger than the rate of deposition. The waves can be natural and caused by the wind, or they could be artificial due to boat wakes and the like. *See also* low-energy coasts

**HYPOXIA:** A condition where oxygen is not present. For shoreline areas, the soils under mangrove forests and salt marshes are hypoxic, which allows peat to form and carbon to be sequestered.

**INDIVIDUAL PERMIT:** A class of permit available from the Florida Department of Environmental Protection (FDEP). This permit is subject to the highest standard of review. The federal (US Army Corps of Engineers [USACE]) analog of this permit class is called a "standard permit."

**INFAUNA:** Invertebrate aquatic animals that burrow in the sediment, such as clams and worms.

**INTERTIDAL AREA:** The area that is above the mean low water (MLW) line and below the mean high water (MHW) line. This area is flooded and exposed once or twice a day.

**INVASIVE SPECIES:** Plants or animals that are not native to an ecosystem *and* that have been shown to cause harm to native species and/or ecosystem function. The Florida Invasive Species Council determines which plant species are invasive in Florida.

**KEYSTONE SPECIES:** A keystone species is a species of plant or animal that has a disproportionately large effect on other species in its natural environment. If it were removed, the ecosystem would change drastically.

**LENTICEL:** A porous tissue arrangement in the bark of woody stems or roots that provides a pathway for the direct exchange of gasses from the air through the bark, which is otherwise impermeable.

**LIVING SHORELINE:** A nature-based approach to coastal erosion control and shoreline protection that uses natural materials and ecological processes to safeguard coastal areas.

**LONGSHORE DRIFT AND LONGSHORE CURRENT:** These are geological processes that result in the transportation of sediments along a coast parallel to the shoreline. The direction of the transport is dependent on the angle of the incoming waves.

**LOW-ENERGY COASTS:** Areas that have low wave and current energy resulting in more sand and sediment being deposited than lost over time. *See also* high-energy coasts

**MARSH VEGETATION:** Herbaceous (non-woody) plants that thrive in wetlands. Most living shorelines deal with the creation of salt marsh. Low marsh is more frequently inundated than the high marsh zone.

**MEAN HIGH WATER (MHW) LINE:** The MHW line on a chart or map represents the intersection of the land with the water surface at the elevation of the average high water.

**MEAN HIGHER-WATER (MHHW) LINE:** The MHHW line on a chart or map represents the intersection of the land with the water surface during king tide events with higher than average water levels.

**MEAN LOW WATER LINE (MLW):** The line on a chart or map that represents the intersection of the land with the water surface at the elevation of average low water.

**MEAN LOWER-WATER (MLLW) LINE:** The MLLW line on a chart or map represents the intersection of the land with the water surface during king tide events with lower than average water levels.

**NATIONWIDE PERMIT (NWP):** A class of permit available from the US Army Corps of Engineers (USACE) that provides expedited permission for projects that meet certain criteria. The most common NWP for living shorelines is NWP 54, but sometimes NWP 13 or NWP 27 will be used to permit living shorelines.

**NATIVE VEGETATION:** Plants that occur naturally in an area, not those brought to an area by humans. The Florida Native Plant Society defines a native plant in Florida as a species that occurred within the state prior to European settlement, based on the best available scientific and historical records.

**NEKTON:** Living organisms that can swim and move independently of currents. Nekton include swimming animals such as fish, squid and similar swimming invertebrates, and marine mammals.

**OSMOSIS:** In plants, osmosis is the movement of a solvent (water) across a semipermeable membrane (a cell wall) to equalize the concentrations of the solvent and the solute.

**OYSTER RECRUITMENT SUBSTRATE:** Structure such as shell or concrete deployed in the intertidal zone to provide oyster attachment opportunities.

**OYSTER REEF BREAKWATERS:** Breakwaters built at intertidal heights (submerged at high tide) out of materials conducive to colonization by oysters.

**PEAT:** Peat is type of soil consisting of partially decomposed organic matter often formed in wet soil with little or no oxygen—a condition called hypoxia. Peat soils can sequester carbon for thousands of years.

**PH:** *See* acidity

**PHOTOSYNTHESIS:** The process where green plants combine carbon dioxide from the air and water from the soil with light from the sun to form simple sugar and release oxygen: $6CO_2 + 6H_2O$ + sun's energy $\rightarrow C_6H_{12}O_6 + 6O_2$. The sugars provide energy for the plant and for everything that eats the plant or absorbs plant material in one way or another through the equal and opposite reaction: respiration.

**PHYTOPLANKTON:** Microscopic algae that photosynthesize at the top of the water column and that are the basis for marine life.

**PLANKTON:** A combination of photosynthetic phytoplankton and the animal zooplankton that feed on them. All plankton move in response to water currents and multiply rapidly when there are more nutrients in the water.

**PLAN-VIEW DRAWING:** A depiction of a living shoreline project from an overhead perspective, including information about the approximate mean low water (MLW) and mean high water (MHW) lines, breakwater/sill placement and spacing, spacing of plants, location of existing structures, location of parcel boundaries, etc.

**POSH (PERVIOUS OYSTER SHELL HABITAT) SUBSTRATE:** POSH substrates are modules of concrete mixed with natural oyster shells that can be used as materials in living shorelines.

**PROJECT DESIGN CRITERIA (PDCS):** Design criteria required for projects to meet the Jacksonville District's Programmatic Biological Opinion in order to qualify for expedited review.

**PROPAGULE:** Propagules are mangrove seedlings that have germinated while still attached to the parent plant.

**PROPRIETARY AUTHORITY:** Florida's State Legislature established a Board of Trustees in 1845 that has proprietary (ownership) authority over sovereign submerged lands (SSL) and their uses; the board trustees were held responsible for ensuring the lands and the associated aquatic resources remained healthy and in abundance for present and future generations. This authority is the reason that even living shorelines that meet the exemption require permission from the state to construct their project on SSL.

**PROP ROOTS:** Red mangroves (*Rhizophora mangle*) have prop roots that arise from lower on the trunk and take root on the bottom. Most of the time, the prop roots are visible above the high water line. *See also* Drop roots

**PSEUDOFECES:** *See* biodeposits

**QUADRAT:** Each of a number of small areas of habitat, typically one square meter each, selected at random to act as samples for assessing the local distribution of plants or animals.

**REEF:** A ridge or shoal of stable material such as rock, coral, or oyster lying beneath the surface of a natural body of water. Many reefs result from natural processes such as layers of oyster recruitment over time, corals building their skeletons, or wave energy eroding rock outcrops.

**REGULATORY AUTHORITY:** Allows an entity of the government, such as the FDEP or the USACE, to limit certain activities on private and publicly owned lands to some specific degree for the greater public good. This authority is the reason there is a permitting process for living shorelines. There are three regulatory agencies that may control living shorelines permitting in Florida: the Florida Department of Environmental Protection (FDEP), the state's five Water Management Districts (WMDs), and the US Army Corps of Engineers (USACE). Note, however, that local governments may have additional permitting requirements.

**RESILIENCE:** The ability of an ecosystem to maintain key functions and processes in the face of stresses or pressures by resisting and then adapting to change.

**RESPIRATION:** The process where organisms (plants, animals, fungi, and bacteria) take in oxygen and absorb sugars produced by green plants through photosynthesis and then release carbon dioxide and water. $C_6H_{12}O_6 + 6O_2 \rightarrow$ energy for life $+ 6CO_2 + 6H_2O$. *See also* photosynthesis

**RETAINED WATERS:** Waters where section 404 permitting authority is retained by the USACE. The USACE retains responsibility for permitting the discharge of dredged or fill material in those waters identified in the Retained Waters List as well as in all waters subject to the ebb and flow of the tide shoreward to their mean high-water mark that are not specifically listed in the Retained Waters List, including wetlands adjacent thereto landward to the administrative boundary.

**RETROFITS FOR HARDENED SHORELINES:** Projects that work to add natural elements to existing armored shorelines. An example would be planters or other structures added to the water side of seawalls or other armored shorelines to create protected spaces for living shoreline plants such as mangroves or marsh grasses to grow. This "softens" the shoreline, and other plants or animals such as oysters may be recruited to the shoreline.

**REVETMENTS:** Sloping structures built on embankments or shorelines, or in front of sea walls, to absorb and dissipate the energy of waves in order to reduce coastal erosion.

**RIPARIAN RIGHTS:** If a waterfront property includes riparian rights, the landowner may access the water, build structures such as docks, and use the water for boating, swimming, or other activities. Waterfront or beachfront properties without riparian rights may not have exclusive use of the water adjacent to the property because public access or some other use has been allowed or granted.

**RIPRAP:** A group of large rocks or other materials used to shore up an embankment and to deter or prevent erosion from waves or currents.

**ROOT FLARE:** At the base of a seed-grown tree, the root flare is a swelling where roots join the trunk; this flare should be visible just above the soil surface. Trees generated by cuttings will not have a true root flare, but the top roots should still be treated in the same way.

**SCARP:** A steep cliff caused by erosion at its base. Also referred to as an escarpment.

**SCARPING:** The formation of a steep or undermined cliff caused by erosion.

**SEDIMENTATION:** The deposition of sediments (from sand to silt) that occurs when particles suspended in water settle out and rest on the bottom. This is usually due to the slowing of the currents or the damping of waves. The heavier particles settle out first.

**SILL:** Material deployed at the base (toe) of a vegetated zone for the purpose of reinforcing and protecting the area from moderate-energy waves and currents. Sills can be constructed out of materials such as shells, concrete, or coconut fiber logs. Sills differ from breakwaters in that they are constructed directly adjacent to the vegetated area.

**SOVEREIGN SUBMERGED LANDS (SSL):** Lands including, but not limited to, tidal lands, islands, sand bars, shallow banks, and lands waterward of the ordinary or mean high water (MHW) line, beneath navigable freshwater or beneath tidally influenced waters, to which the State of Florida acquired title on March 3, 1845, by virtue of statehood, and which have not heretofore been conveyed or alienated.

**STAGGERED DESIGNS:** A breakwater configuration that relies on parallel, offset rows of proposed structures (breakwater, gabion, etc.). This design can reduce the funneling of wave energy landward of the structures and provide a greater level of property protection.

**STATE PROGRAMMATIC GENERAL PERMIT (SPGP):** An agreement between FDEP and USACE that allows the state to issue a joint permit that covers both state and federal requirements in cases that meet the criteria laid out in the agreement.

**STORMWATER RUNOFF:** Rainwater that flows over land or impervious surfaces (paved streets, parking lots, rooftops) and does not soak into the ground. Runoff can carry pollutants from the land into waterways. Stormwater runoff from the uplands can also cause erosion and should be considered in living shoreline project designs.

**SUBMERGED AQUATIC VEGETATION (SAV):** Grasses that grow to the surface of—but do not emerge from—shallow water. This includes seagrasses and freshwater aquatic grasses.

**SUBTIDAL:** The area of the shoreline below the mean low water (MLW) line; subtidal describes the area that would be underwater almost all the time. *See also* intertidal

**SUFFICIENT UPLAND INTEREST:** A legal term referring to ensuring that a permit applicant has ownership (or permission from owners) of riparian areas adjacent to projects that will occur along or near riparian areas. Acceptable documentation includes a warranty deed; a certificate of title issued by a clerk of the court; a lease; an easement; or a condominium, homeowners, or similar association document. Documents must clearly demonstrate that the holder has control and interest in the riparian uplands adjacent to the project area and the riparian rights necessary to conduct the proposed activity.

**TOE:** When referring to natural shorelines or living shorelines, the toe is a narrow region around the mean low water (MLW) and the lower end of the intertidal zone, which is also where the marsh plants begin to grow. Living shorelines are often designed to protect the "toe" of the marsh plants.

**TRANSECT:** A transect is a straight line that cuts through a natural landscape so that standardized observations and measurements can be made. Transects help ecologists measure, make observations, and record data across a monitored plot. Transect observations record the complexity of a natural environment in a way that can be tracked over time and compared to other areas.

**TRANSPIRATION:** The process in vascular plants where water that is absorbed by osmosis through their roots travels through the plant and then evaporates into the air through the stomata (pores) on the leaves.

**TUNICATE:** A primitive marine animal, commonly called a sea-squirt, that spends most of its life attached to docks, rocks, or the undersides of boats.

**TURBIDITY:** The measure of suspended solids, which influences water clarity. Murky water has high turbidity.

**UPLANDS:** Florida's Division of State Lands defines uplands as those lands above the mean high water (MHW) line (or ordinary high water line for freshwater habitats).

**US ARMY CORPS OF ENGINEERS (USACE):** There are three permitting branches: The North Branch includes Pensacola, Panama City, Jacksonville, and Cocoa Permit Sections. The West Branch includes Tampa and Ft. Myers. The South Branch includes Palm Beach Gardens, Antilles-Miami Section, and the Keys Section.

**VASCULAR PLANTS:** Plants that have tubes or veins, called xylem, that carry water to parts of the plant that are not in contact with moisture, mostly the parts above ground. Other veins, called phloem, carry sugars and other nutrients both up and down to all parts of the plant. The phloem includes flaps between the cells to control the direction of the flow.

**VERIFICATION OF EXEMPTION:** A letter issued by the Florida Department of Environmental Protection (FDEP) verifying that a proposed project meets exemption criteria and can proceed without additional permitting.

**VICINITY/LOCATION MAP:** A map that should provide sufficient detail to allow one to find the subject parcel and include the names of major streets and roadways, the section/township/range of the project, the latitude and longitude of the project in decimal degree format, and street address of the project (if available).

**VIVIPAROUS GERMINATION, OR VIVIPARITY:** Born alive. For plants, this refers to seeds that germinate while still attached to parent plants.

**WATER MANAGEMENT DISTRICTS (WMDS):** There are 5 WMD sections: Northeast, Suwannee River, St. Johns River, Southwest, and South Florida. A map of the offices is located at https://floridadep.gov/water-policy/water-policy/content/water-management-districts.

**WATER QUALITY CERTIFICATION (WQC):** Section 401 of the Clean

Water Act requires any applicant for a federal license or permit to conduct any activity which results in a discharge into Waters of the United States (WOTUS) to provide the federal agency with certification from the certifying authority that the discharge will comply with applicable water quality standards.

**WATERWARD TOE:** The bottom-most portion of the offshore side of a structure such as a sill or breakwater.

**WETLAND:** An area that is inundated or saturated by surface water or groundwater at a frequency and a duration sufficient to support, and under normal circumstances do support, a prevalence of vegetation typically adapted for life in saturated soils.

**WRACK LINE:** This is the area on the shoreline where organic material and other debris is deposited after a high water event.

**ZOOPLANKTON:** Small nonphotosynthetic animal organisms that feed on the phytoplankton and other small organisms in the water. *See also* plankton

# Index

Page numbers in *italics* indicate illustrations.

**SAVANNA BARRY** serves as a regional specialized extension agent at the University of Florida / Institute of Food and Agricultural Sciences (UF/IFAS) Nature Coast Biological Station in Cedar Key, Florida. Her work focuses on the intersection of environmental management and applied research and supports community-driven coastal science. Savanna grew up on a small farm in central Virginia and discovered her interest in marine ecology during family vacations to small fishing towns along the Chesapeake Bay. After earning her BS in Biology from the University of Virginia, Savanna split her time between Gainesville, Florida, and Little Cayman Island to earn her MS in Fisheries and Aquatic Sciences through the University of Florida. While in Little Cayman, Savanna also completed her SCUBA Divemaster certification. After earning her MS, Savanna received a PhD from the University of Florida by studying the influence of water quality on the ecosystem function of seagrass meadows. Savanna enjoys traveling, cooking, reading, birding, attending Gator games, and spending time with her family and fur babies.

**VINCENT ENCOMIO** is a UF/IFAS and Florida Sea Grant extension agent based on the Treasure Coast of Florida, serving Martin and St. Lucie Counties. Vincent received his bachelor's degree in biology from the University of California, San Diego. He continued his studies in marine biology and marine science, receiving his master's from San Francisco State University and his PhD from the Virginia Institute of Marine Science at the College of William and Mary. His extension education program is related to coastal habitat restoration and water quality. He brings science-based information to the public through stormwater education; citizen-science programs in shoreline, water quality and HAB (harmful algal bloom) monitoring; and education on estuarine habitats and living shorelines through the Florida Master Naturalist Program and the Living Shorelines Training for Marine Contractors. In his spare time, Vincent enjoys spending time with his family at the beach and annual family scalloping trips to Florida's Big Bend area. He is an avid surfer and enjoys snorkeling, gardening, and walking his dogs when the waves are flat.

**MARJORIE SHROPSHIRE** was born in Miami, Florida, and grew up making regular forays into the Everglades, the Florida Keys, and the lower southwest Florida coast. She began painting at an early age and became particularly interested in depicting natural history topics. After earning a BFA from the University of Miami, Coral Gables, she embarked on a career as a creative director, and in 2006, established her own design practice, Shark River Creative, which focuses on Florida's environment. She edits and produces the Florida Native Plant Society's quarterly magazine, *Palmetto,* and she conducts outreach programs centering on Florida's native plants and how they affect water quality and sustainability. She is a coauthor with Ginny Stibolt on *A Step-by-Step Guide to a Florida Native Yard* (2018), and she has illustrated several other books on botanical topics. Her lifelong interest in the marine environment led Marjorie to earn an MFA in Design and Visual Communications from the University of Florida, where her research centered around using design to connect homeowners and practitioners interested in installing living shorelines. In her free time, she enjoys botanizing, sailing, adventure travel, volunteering on endangered plant restoration projects, and teaching others how to closely observe the natural world through nature journaling.

**GINNY STIBOLT** is a lifelong gardener, award-winning garden writer, and an advocate for earth-friendly practices. She earned an MS in botany at the University of Maryland. When she and her husband moved to northeast Florida in 2004, she discovered that gardening in Florida is different from her previous experiences in New England and the Mid-Atlantic regions, and since then, she has been writing about her adventures in Florida gardening. In 2006, after joining the Florida Native Plant Society, she changed her landscaping style to include more native plants and more natural areas in her yard. She wrote *Sustainable Gardening for Florida* (2009), *Organic Methods for Vegetable Gardening in Florida* with Melissa Markham Contreras (2013), *The Art of Maintaining a Florida Native Landscape* (2015), *A Step-by-Step Guide to a Florida Native Yard* (with Marjorie Shropshire) (2018), and *Adventures of a Transplanted Gardener* (2022). Also, she cowrote the award-winning *Climate-Wise Landscaping: Practical Actions for a Sustainable Future* with Sue Reed (2018). She has written hundreds of articles—both in print and online—for various organizations. In addition, she manages the Sustainable Gardening for Florida Facebook page.